Westfield Middle School
Media Center

Extraterrestrials

A FIELD GUIDE FOR EARTHLINGS

TERENCE DICKINSON AND ADOLF SCHALLER

ILLUSTRATIONS BY ADOLF SCHALLER

CAMDEN HOUSE

Canadian Cataloguing in Publication Data

Dickinson, Terence
  Extraterrestrials : a field guide for earthlings

Includes index.
ISBN 0-921820-86-0 (bound)
ISBN 0-921820-87-9 (pbk.)

1. Life on other planets - Juvenile literature.
I. Schaller, Adolf.   II. Title.

QB54.D53 1994   j574.9992   C94-931194-4

Published by Camden House Publishing
(a division of Telemedia Communications Inc.)

Camden House Publishing
7 Queen Victoria Road
Camden East, Ontario K0K 1J0

Camden House Publishing
Box 766
Buffalo, New York 14240-0766

Printed and distributed under exclusive licence
from Telemedia Communications Inc. by
Firefly Books
250 Sparks Avenue
Willowdale, Ontario
Canada M2H 2S4

Firefly Books (U.S.) Inc.
P.O. Box 1338
Ellicott Station
Buffalo, New York 14205

Printed and bound in Canada by
Metropole Litho Inc.
St. Bruno de Montarville, Quebec

Colour separations by
Hadwen Imaging Technologies
Ottawa, Ontario

Design by
Linda J. Menyes

ACKNOWLEDGMENTS

To Tracy Read and Linda Menyes, the best book editor and designer in this part of the universe.   –T.D.

To my nephews Eric and Ryan, who by virtue of their youth know, perfectly well, an alien world when they see one.   –A.S.

The production of a book is rarely a one-man or one-woman proposition, and this one is no exception. Far from it. Not only did my collaborator, Adolf Schaller, provide many stunning original paintings, but his copious notes accompanying them helped shape the book into its final form. Designer Linda Menyes produced another beautiful book (our third collaboration) to display the pictures and words. Tracy Read, my editor at Camden House, offered many insightful comments that improved the final product. Jody Morgan took on the tough job of tracking down the movie stills we needed; Catherine DeLury supplied her unerring proofreading skills; and science fiction expert Christine Kulyk offered many useful suggestions and always managed to dig up obscure but essential facts. As always, my largest thanks goes to my partner and world-class copy editor and production coordinator Susan Dickinson.
                                    —Terence Dickinson

CONTENTS

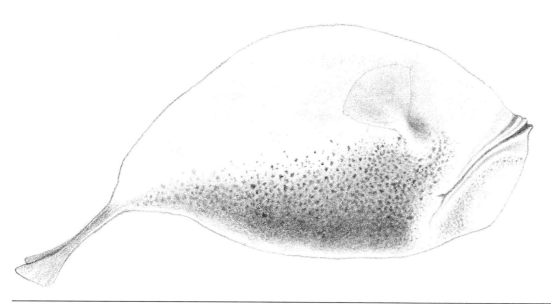

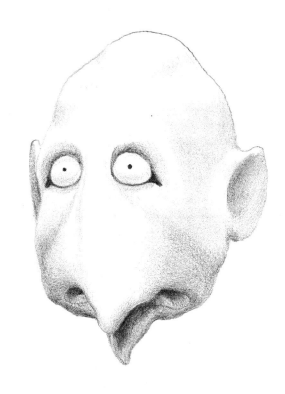

ARE WE ALONE?

◆ Earth is the only place in the universe where life exists—as far as we know. But our knowledge of worlds beyond our own is limited to the eight other planets in our solar system, their five dozen moons and some asteroids and comets. None of them appear to support life, but what about the countless planets that might orbit around other stars? Astronomers estimate that there are at least a billion trillion stars in the universe. They could provide a vast array of interesting homes for extra-terrestrial life.

What will *they* look like? Extraterrestrials, that is. We've seen lots of aliens on television and in the movies—everything from Mr. Spock to talking rocks.

Some of those television and movie aliens look exactly like us. Some are just a bit different, with weird skin colour, pointy ears or distorted facial features. Others are bizarre, like the giant walruslike Jabba the Hutt in the *Star Wars* movie *Return of the Jedi.*

But have you ever wondered whether these aliens actually make sense? What do they eat? What is their home planet like? What type of air do they breathe?

In this book, we venture beyond the aliens we know from television and the movies and try to be a little more scientific in our approach as we explore what extraterrestrials might be like—if they exist. We examine the possibilities for life as we know it and travel into the shadowy realm of life as we don't know it. We have tried to blend science and imagination to depict life on other planets. It's a voyage into the possible, if not the probable.

One thing we do know when we look around us at the other planets in our own solar system is that we don't find worlds with surface conditions anything like the Earth's. Not a drop of water exists on our moon. The nearest planet, Venus, has a surface hot enough to melt lead, even at its poles. Our other nearby planetary neighbour, Mars— the planet most like Earth—is a frigid desert world. Mars and Venus are the planets where astronomers expected to find conditions similar to those on Earth.

In fact, the more we learn about other worlds, the more we realize that an enormous variety of conditions exists out there. Although only a small fraction of those environments could support life as we know it, the universe, as we shall see, is a very big place.

Imagine that each star in the known universe is represented by a single grain of sand. A thimble would hold all the stars visible on a clear, dark summer night. A large construction wheelbarrow would contain the Milky Way, the galaxy in which our sun resides. But to demonstrate the immense number of stars in the universe, we need a freight train with hopper cars filled with sand. As the train begins to speed by us at a level crossing, we count the cars while we wait. The cars roar past, one per second. We would have to keep count 24 hours a day for *three years* before the universe train had completed its pass.

If only one star in a billion is parent to a planet with life, then at least one trillion planets in the universe harbour living matter. Suppose that during the history of the Milky Way Galaxy, a few hundred other intelligent technological civilizations arose on places much like Earth. What would these civilizations be like? It's a powerful question—one that carries the mind to limitless speculation.

◆ Top: Just about everyone has, at one time or another, been impressed by a view of the night sky filled with countless twinkling stars. Unfortunately, many of us see the sky properly only during vacations that take us far from city lights. But when the stars shine forth, thoughts of life on other worlds soon follow.

◆ Bottom: Each star is another sun basically like our own. A typical galaxy, similar to the one shown here, contains more than 100 billion stars. Photographs taken with large telescopes have revealed billions of galaxies, and there are undoubtedly billions more yet to be discovered.

One of the earliest motion pictures made—a 1902 production from France called *A Trip to the Moon*—featured the first extraterrestrials ever seen in the movies. These were Selenites, or moon men, created by the film's director, Georges Méliès.

Today, this 16-minute film looks pretty silly, but in the early 20th century, audiences found it hugely entertaining. It begins with chorus girls escorting a crew of astronauts into a bullet-shaped space capsule. The capsule is then loaded into a giant cannon that blasts the astronauts to the moon. There, they encounter the Selenites, fight them off and narrowly escape to return to Earth.

The idea that we might someday travel to the moon and meet creatures which inhabit that world did not originate with Méliès. He based his film on Jules Verne's story *From the Earth to the Moon* (published in 1865) and H.G. Wells' book *The First Men in the Moon* (1901). Verne and Wells were two of the first

Real Moon Men

Between 1969 and 1972, a dozen explorers from Earth—the Apollo astronauts—walked on the moon. There, they discovered an airless, inhospitable world of rock and dust. Analysis of the moon rocks confirmed that no liquid water has ever existed on the moon. It is a world that has never supported life.

writers of what we now call science fiction. Apparently, Méliès was a big fan of theirs and used parts of both works for his film.

In Verne's tale, the explorers are launched into space by a colossal cannon, but they miss the moon, orbit it and return to Earth. The astronauts in Wells' story reach the moon and find it inhabited by Selenites, which resemble large insects and live underground. Méliès' Selenites were based on those in Wells' novel. In the film, they were played by human acrobats and can be seen flipping over and around the curious Earthlings.

Méliès used a formula that still works today: He focused on entertainment rather than realism. He knew his audiences wanted adventure, a fast-paced story, some pretty faces and something they had never seen before: extraterrestrials.

Amazingly, no other films about extraterrestrials were made for almost 50 years, with the exception of a few Buck Rogers and Flash Gordon serials that were produced to be screened between double features at Saturday movie matinees. (In the era before television, kids regularly spent Saturday afternoons at the local cinema watching black-and-white Westerns or adventure movies.) With titles such as *Flash Gordon Conquers the Universe* and aliens that looked like adults in Halloween costumes, these low-budget action serials were so bad,

nobody took them seriously. But this was understandable in one way. It was a time when travelling in space was considered by almost everyone as something that would happen in the remote future.

By the late 1940s, however, science fiction writing was reaching maturity. Authors such as Arthur C. Clarke and Robert Heinlein were creating intriguing stories about alien civilizations and extraterrestrial visits to Earth. Rockets had been invented and were used during World War II (1939-1945). A few experts were beginning to suggest that we might be able to get to the moon before the year 2000.

By 1950, Hollywood finally started paying attention. That year, *Destination Moon* was released, the first serious film about travel to another world. In it, astronauts fly to the moon by rocket, explore its rugged surface and return to Earth. No extraterrestrials were sighted, but they would soon follow.

◆ Top right: The first extraterrestrials seen in a motion picture were clawed humanlike creatures designed by pioneer moviemaker Georges Méliès for his 1902 film *A Trip to the Moon*. They were actually acrobats in costume, but audiences loved them and packed the movie houses to catch a glimpse of the aliens.
◆ Top left: Explorers examine the lunar landscape in a scene prepared for *A Trip to the Moon*.

◆ In this scene from *A Trip to the
Moon*, an astronaut dangles precari-
ously as a Selenite (moon man) tries
to push his spaceship off a cliff on the
moon. The movie was based on stories
by Jules Verne and H.G. Wells.

Many of us had our first experience with extraterrestrials at the movies or on television, perhaps in one of the *Star Wars* films or a rerun of *Star Trek*. Alien creatures abound in recent Hollywood productions, but whether these creatures are created for the big or small screens, you've probably noticed that most Hollywood extraterrestrials look like humans except for the head. There are two main reasons for this.

The first is practical: It is much easier to stick a weird latex head on an actor than it is to design and build a whole alien. Unless there is somebody inside, the creature must be either filled with pistons and pulleys to produce body motion or operated remotely with wires like a puppet, not unlike the way in which Kermit the Frog and the other friendly Muppets are controlled.

The second reason is a little more complicated. Humans are very sensitive to distortions in faces. It doesn't require much manipulation to the features we are familiar with before we see a monster or an alien instead of another human like ourselves. That is why the Hollywood special-effects departments concentrate on faces and heads—bulges here and

◆ Top: Mr. Spock of the original *Star Trek* television series sports the pointy ears and eyebrows that signify he is half extraterrestrial (his father lived on a planet called Vulcan). Spock quickly became the most popular

Star Trek character. But this was mostly because of actor Leonard Nimoy's convincing portrayal of Spock's extraterrestrial traits, rather than the effectiveness of elaborate Hollywood makeup.

◆ Bottom: These extraterrestrial humanoid astronauts were members of the Rebel Alliance in a scene from *Return of the Jedi*, one of the *Star Wars* films.

there, horns, bizarre hairdos, and so on.

Mr. Spock, from the original *Star Trek*, is perhaps the best-known example of an almost-human alien. Portrayed by actor Leonard Nimoy, Spock is half human and half Vulcan. Apart from his elongated, pointed ears, he could easily pass for a human. Yet Nimoy's portrayal of Spock's nonemotional Vulcan temperament convinces us that he is not.

The original *Star Trek* and its successor, *Star Trek: The Next Generation*, and its spin-offs, *Star Trek: Deep Space Nine* and *Star Trek: Voyager*, feature a whole series of humanlike extraterrestrials. Many are actors in latex headgear or masks. Some look exactly like humans, which, as we will see later, is extremely unlikely.

Hollywood has always been fascinated by monsters, but extraterrestrials were rarely portrayed until the 1950s, when the movie studios began to churn out dozens of science fiction films populated by various creatures from other planets. The sudden interest in extraterrestrials was sparked in part by the first widespread UFO (unidentified flying object) sightings and by the fact that scientists were conducting experiments with rockets which might someday carry people into space. Combined with the ominous discovery of atomic power, these developments were more than enough to spawn a flock of (mostly awful) science fiction films.

Even the names of the 1950s films suggest that Hollywood did not have the slightest intention of taking extraterrestrials seriously. Here are some examples: *I Married a Monster From Outer Space* (1958), *Invasion of the Saucer Men* (1957), *It Came From Outer Space* (1953), *The Thing From Another World* (1951), *The Brain From Planet Arous* (1958), *Earth vs. the Flying Saucers* (1956), *It! The Terror From Beyond Space* (1958) and *Devil Girl From Mars* (1954). They were all just as bad as—and sometimes even worse than—they sound.

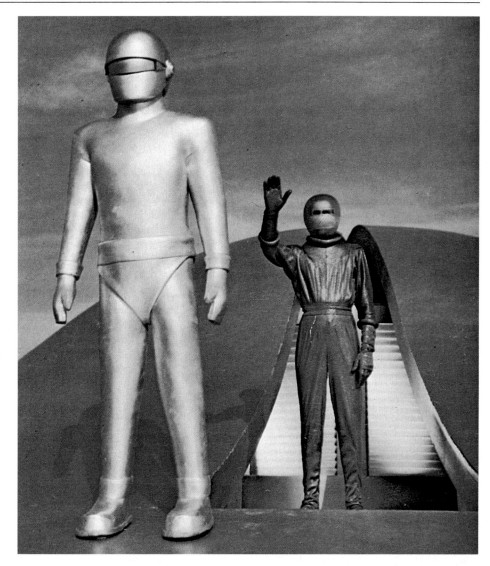

There were exceptions, of course. One of the first of the 1950s science fiction films was also one of the best. *The Day the Earth Stood Still* (1951) features Klaatu, an alien who looks exactly like a human except he is quietly but clearly smarter than we are. By taking human form, Klaatu is readily accepted on Earth. (This neat trick also saved the movie's producers from spending a pile of money on alien costumes and makeup.)

Klaatu's intentions are peaceful but persuasive. He brings with him a huge robot named Gort. The galactic civilizations dispatch such robots as a police force to control aggressive creatures like Earthlings. Gort demonstrates his awesome powers by melting tanks and other weapons, but the entire movie is intelligently done and remains a classic.

The Day the Earth Stood Still was the first Hollywood movie to treat the subject of extraterrestrial life seriously. It is available on videocassette and is well worth seeing.

◆ A scene from the classic 1951 film *The Day the Earth Stood Still* shows the extraterrestrial Klaatu greeting Earthlings from the deck of a flying saucer. Later, Klaatu announces that the giant robot, Gort, is a member of a galactic police force established to prevent aggressive creatures like humans from spreading their warlike ways to other planets. This was the first science fiction film to treat the idea of extraterrestrials seriously.

Science fiction author Arthur C. Clarke tells the story of how he and moviemaker Stanley Kubrick were struggling with the ending of the classic 1968 science fiction film *2001: A Space Odyssey*. They had decided that Bowman, the last surviving crew member of the ill-fated Jupiter expedition, would be transported through space to meet the extraterrestrials that had left a signalling device near Jupiter. But what would they look like?

"We're not going to use guys from central casting wearing rubber masks," argued Kubrick. "There's been too much of that. It's too unimaginative. We have to do something different."

And they did. After mulling over the problem for months, they decided not to show the extraterrestrials at all. Yet during the movie, the audience realizes the intellectual and technological power of the aliens and how incredibly advanced they must be. The result is a spectacular film that many consider to be the best science fiction movie ever made.

Despite the success of Kubrick's landmark work, no one else has produced anything like it. In fact, the trend has been to show as many weird aliens as possible. Certainly this was the thinking behind the menagerie of aliens in the *Star Wars* trilogy—*Star Wars*, *The Empire Strikes Back* and *Return of the Jedi*—which are among the top money-making movies of all time.

One of the most memorable scenes in *Star Wars* takes place in the cantina, a sleazy bar at the Mos Eisley Spaceport, described in the movie as a "wretched hive of scum and villainy," on the fictional planet Tatooine. It's an interplanetary crossroads, and the place is filled with bizarre aliens the likes of which had never been seen on movie screens before. This sequence alone was worth the price of admission when the film was released in 1977, and millions of people flocked to the theatres to see the creatures.

The next two movies in the series had even more detailed portrayals of extraterrestrials, especially *Return of the Jedi*, in which the walruslike Jabba the Hutt is the extraterrestrial version of a mobster and drug dealer. From his hideout somewhere on Tatooine, Jabba controls an interplanetary smuggling operation. After botching a job for Jabba, Han Solo, the movie's hero, is imprisoned in Jabba's headquarters. The scenes of Jabba and his henchmen—a grotesque collection of unsavoury alien creatures—are probably the most imaginative and entertaining of their kind.

To create Jabba's complex motions, two special-effects technicians sat inside the monster's body, pushing rods and grasping levers. Their movements had to be precisely coordinated to make Jabba's lips, arms, eyes and body move the right way at the right time. The result is a compelling portrayal of a being that seems to be truly out of this world.

◆ Top: An extraterrestrial from the television series *Star Trek: Deep Space Nine*, a programme which relies heavily on head masks and detailed facial makeup for that alien look.
◆ Bottom: The scenes in *Return of the Jedi* that take place in Jabba the Hutt's hideaway include the most amazing collection of extraterrestrials ever seen on film. If nothing else, they suggest that drooling is one of the primary activities of alien creatures.

But the *Star Wars* movies were never intended to be science fiction. They are pure fantasy escapism—a good-guys-versus-bad-guys adventure that takes place on other worlds. Sometimes called space operas, such stories, if done well, can be hugely entertaining, but they don't tell us much about interstellar travel or life on other worlds.

One thing these films have demonstrated is that extraterrestrials are immensely popular. But in many ways, the extraterrestrials and their environments as depicted in such movies are flawed. For instance, all the extraterrestrials seem to manage just fine breathing the same air.

On the other hand, the science fiction movie monsters of the 1950s were far worse: drooling bug-eyed creatures that were so hokey, movie companies have not bothered to convert most of these films to videocassettes. They are destined to fade into obscurity—an appropriate fate for the majority of them.

◆ Left: The realistic body motions of this extraterrestrial from *Return of the Jedi* were produced using the same puppetlike techniques that give "life" to the Muppets.
◆ Right: This classic bug-eyed monster from the 1955 film *This Island Earth* was a "mutant" from the planet Metaluna. Of course, it's an actor in a rubber suit, but this extraterrestrial succeeded in scaring moviegoers.

During a meeting of the American Association for the Advancement of Science several years ago, news reporters were asked to fill out a questionnaire. One of the questions was: "What, in your opinion, would be the biggest science story of all time?" The majority of the reporters said that the ultimate headline would be the discovery of intelligent life elsewhere in the universe.

Such an event would forever alter our perception of humanity's place in the universe. But so far, although we continue to hear incredible accounts of encounters with UFOs and their alien occupants, there is no real proof that anybody is out there. Could extraterrestrials be right here, in the Earth's vicinity?

There are more than 200 cases involving people who claim to have been taken, against their will, into alien spaceships. There, they were reportedly examined by humanoid creatures before being released. Many "abductees" say their recollections of the encounter were just vague memories.

Are UFOs Real?

Taken aboard a ship near Trinidad in 1958, this picture is typical of hundreds of UFO photographs published since the term "flying saucer" was coined in 1947. Many have been shown to be reflections in the camera lens or simply fakes. We need more concrete evidence than fuzzy pictures to prove that we are being visited by extraterrestrials.

Eventually, however, nightmares about the unsettling experience prompted them to seek help. The details of the incident are often uncovered only through hypnosis.

Several motion pictures have been made about these sensational abduction cases. *The UFO Incident* tells the story of the most famous abductees, Betty and Barney Hill of Portsmouth, New Hampshire. The abduction of Arizona lumberjack Travis Walton is presented in *Fire in the Sky*. As entertaining as these movies are, they merely repeat the stories told by the abductees, rather than offer new facts.

Yet the idea that someone has been captured by aliens, examined and then released is so sensational that any proof of it would be eagerly examined by researchers. Unfortunately, there is no proof. These are just stories. The abductees honestly believe something unusual happened to them, but just what it was remains a mystery to everybody. If extraterrestrials are randomly examining Earthlings, we are going to need more than these isolated tales of abduction to prove it.

The same applies to reports of strangely moving objects in the sky. Although pictures of UFOs may seem convincing, not one UFO photograph has shown something that is unquestionably not of this Earth.

Times change. These days, there isn't much media coverage about UFOs, apart from the completely outrageous stories screaming from the front pages of tabloid newspapers at supermarket checkout counters. UFOs may one day become a news worthy subject again, but for now at least, most people have lost interest.

Curiosity about UFOs was much higher in the 1950s, 1960s and early 1970s. That's what made moviemaker Steven Spielberg interested in producing a big-budget movie about UFOs and our first contact with extraterrestrials. Spielberg's film, *Close Encounters of the Third Kind,* was released in 1977 and became one of the top money-making movies up to that time.

In the film, actor Richard Dreyfuss plays a man obsessed with finding out what happened to him when he had a close encounter with a flying saucer. He manages to be on hand when the mother ship lands and the extraterrestrials make their first attempt to interact with Earthlings.

Close Encounters of the Third Kind portrays the aliens as childlike in stature, completely harmless and eager to share knowledge with us. This is in sharp contrast to the usual UFO movie theme, which has the creatures from other worlds intent on taking over our minds, our bodies or our planet —or all three.

◆ Top left: This creature from the film *Fire in the Sky* is based on descriptions from people who say they have been taken inside alien spaceships against their will and examined by extraterrestrials. Not one "abductee" has ever supplied a single piece of solid evidence to prove that the abduction actually took place.
◆ Top right: Sketch of the alien features most often described by abductees.

◆ Extraterrestrials greet an Earthling in the final scene of the film *Close Encounters of the Third Kind*. How would a civilization from another world choose to communicate with us? That question is one of the great puzzles of modern science. Would members of the alien civilization simply choose their time and land somewhere for all to see? Would they secretly abduct people for brief examinations, like wildlife game wardens? Or would they contact us by signalling from afar? At the conclusion of this book, we will examine the possibilities further.

Mars has always been more mysterious than any other object in the night sky. When seen among the stars, this planet has a pale reddish or rusty tint, which makes it the celestial object closest in colour to blood. That is not the type of association we would make today, but thousands of years ago, our ancestors equated Mars with war, death and destruction.

Not long after the invention of the telescope in 1609, astronomers noticed that Mars was the only planet which resembles Earth. It has polar caps that expand and retreat with the Martian seasons and pinkish deserts splotched with curious dark regions. Even the occasional cloud obscures the surface. The idea that such a world might be inhabited was accepted by many astronomers hundreds of years ago.

In the late 1600s, Dutch astronomer Christiaan Huygens wrote that he had no doubt Mars and other planets "have creatures to stare and wonder at the works of Nature."

Mars is a difficult object to study. A fairly

◆ Fascinated by the possibility of proving that Martians exist, Percival Lowell built his own observatory in 1894 to study the red planet. He equipped the observatory with one of the largest telescopes in the world at that time. Today, more than a century later, the telescope is still in use on Mars Hill, in Flagstaff, Arizona.

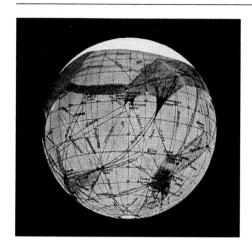

large telescope is necessary to reveal anything at all. Really good views are possible for only two or three months every two years, when Earth and Mars are closest. During just such a close approach in 1877, long, thin, straight lines were observed on Mars by Italian astronomer Giovanni Schiaparelli. He dubbed them *canali*, which in Italian refers to channels or grooves. They soon became known in English as the canals of Mars.

The idea of life on Mars fascinated American astronomer Percival Lowell, and he decided to do something about it. In 1894, he built an observatory in Arizona specifically to look for proof that a Martian civilization had constructed the canals. Lowell's telescope was larger than Schiaparelli's, and the more he looked at Mars, the more canals he saw. His maps of Mars show hundreds.

According to Lowell, the canals were built by a Martian civilization in an effort to preserve a dwindling water supply on a planet that was three-quarters desert. He thought that to be able to construct a system of canals spanning their entire planet, the Martians must be more advanced than Earthlings. Lowell was convinced that the canals were the first proof of the existence of intelligent extraterrestrials.

Although most astronomers disagreed with Lowell's canal theory, the general public found it compelling. People travelled great distances to hear Lowell speak about Mars, and his books sold briskly.

It took nearly a century to prove beyond any doubt that the canals do not exist. The final proof came in 1972, when the American space probe Mariner 9 mapped the entire planet and didn't discover a single canal. The mysterious dark regions are simply areas of darker rock.

Why were Lowell and his followers so certain about what they thought they saw on the surface of Mars? As we now know, the brain tends to connect fine detail into linear, or stringlike, features. (The same effect can be seen if you toss a handful of seeds or crumbs onto a flat surface. Your brain immediately tries to connect the individual pieces into some kind of pattern.) Lowell was forcing his eyes, and his telescope, beyond their limits. Where delicate shading or a series of dark splotches were just at the edge of vision, his brain saw lines.

Something else was going on in the minds of these astronomers. They *wanted* to find another Earthlike world. Anything that looked familiar to them as they watched the shimmering image of a planet through a telescope could trigger ideas and theories. And because Mars looked more like Earth than any other planet did, it became the focus of the greatest number of theories about extraterrestrials.

Even as recently as 1965, some of Lowell's canals were still on the official maps of Mars issued by NASA. The 1965 NASA *Sourcebook of Space Sciences* stated that "most astronomers would probably agree that there are apparently linear markings…of considerable length on the surface of Mars." But there aren't.

War With the Martians

Around the same time that Percival Lowell developed his theory of the Martian canals, H.G. Wells published *The War of the Worlds*, an exciting novel about Martians leaving their dying planet to invade Earth. This illustration of the Martians' fighting machines appeared in an early edition of Wells' book. The machines seemed invincible until—just in time to save humanity—the Martians were defeated not by our weapons but by simple bacteria.

In 1938, *The War of the Worlds* was scripted for a one-hour radio drama and broadcast throughout the United States the night before Halloween. Anyone who tuned in partway through the show heard fake on-the-scene news reports of Martian fighting machines invading New Jersey and blasting everything in sight with a "heat ray." In an early example of the power of the media, thousands of listeners actually believed what they were hearing and panicked. Fortunately, no one was killed trying to escape the fictional Martian attack. They must have felt pretty silly when the commercials came on.

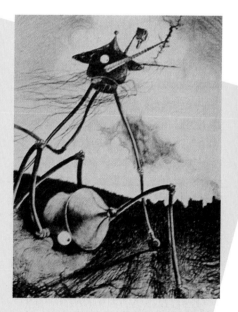

In 1953, *The War of the Worlds* was turned into a Hollywood movie. This time, there wasn't any panic in the streets. Instead, the theatres were filled with enthusiastic audiences.

◆ Above: Percival Lowell's 1905 chart of Mars shows dozens of canals crisscrossing the planet. Lowell was convinced that the canals were built by an intelligent civilization. His theory was widely accepted by the public, but most astronomers said they couldn't see any canals on Mars. Space probes to Mars eventually proved that they were right and Lowell was wrong.

Around midnight on the night of August 22, 1924, Mars was closer to Earth than it had been since well before the telescope was invented. The newspapers were full of speculation. If ever the Martians were going to attempt communication with Earth, this would be the night.

In North America and elsewhere, military radio transmitters were silenced. Many radio stations agreed to end their broadcasts early in order to keep the airwaves open for signals. The most sensitive radio receivers were at the ready. Standing by to translate any alien message was William F. Friedman, chief of the U.S. Army Signal Corps' code division, the same man who later cracked a secret Japanese code just before the attack on Pearl Harbor.

Of course, nothing was heard but static. Yet at the time, it was serious business. People really believed that intelligent life might exist on our neighbour world.

The most shocking revelation about Mars

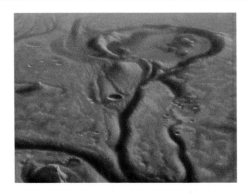

came in July 1965, when the American space probe Mariner 4 passed within 10,000 kilometres of the red planet and returned 19 pictures that revealed a barren, cratered landscape. Mars, it appeared, was not like Earth at all. It resembled the moon.

As Mariner 4 passed behind Mars, distortions in its radio signals as they travelled through the Martian atmosphere allowed scientists to measure the atmosphere's pressure and density. The Martian atmosphere

proved to be less than 1 percent as dense as the Earth's—far too thin to breathe. Even more discouraging was the fact that the space probe detected just a trace of water vapour.

Nothing more promising was seen when Mariners 6 and 7 took more pictures in 1969. The search for life on Mars seemed to be over.

Then came Mariner 9, which went into orbit around Mars in 1971. Previous probes had seen only 10 percent of the planet—all craters and windswept plains. But elsewhere on Mars, Mariner 9's cameras revealed vast chains of volcanoes, some of which would dwarf Mount Everest. An astonishing rift valley, torn open by geologic stress, stretches for thousands of kilometres. The Grand Canyon could be tucked unnoticed into one of its tributary arms.

But the most amazing of Mariner 9's revelations was the dried riverbeds. Apparently, Mars had once had flowing liquid water on its surface. Maybe Percival Lowell had been right about one thing: At some point, Mars had dried up and become less suited for life.

While the planet seemed unlikely to harbour advanced life forms, these discoveries again fired excitement about finding life on Mars. Since hardy microbes such as bacteria thrive under the worst conditions on Earth, perhaps simple life forms on Mars had adapted to the gradual loss of water.

The search for life on the surface of Mars was conducted by two American Viking probes, each about the size of a small car,

The 'Face' on Mars

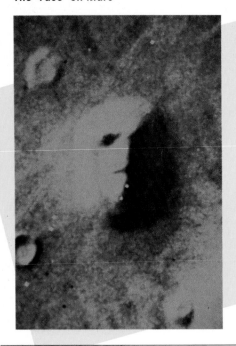

In 1976, two American Viking Orbiter spacecraft went into orbit around Mars. Over the next two years, they took thousands of pictures of the red planet's rocky surface.

One of the Viking Orbiter pictures has become famous. It shows a feature about the size of 20 city blocks that resembles a human face staring straight up, like a colossal carving. The formation is made of rock. We know this because it is the same colour as its surroundings.

Some people have claimed that the "face" was built by Martians long ago to attract our attention when we finally reached Mars. When considering the merits of this claim, remember that the Viking Orbiters took 60,000 pictures of Mars showing millions of landforms. It is probably not surprising that one of them has a familiar shape. When more details of this feature are revealed by future space probes, it will be obvious that extraterrestrials had nothing to do with it.

◆ Top left: There is no liquid water on Mars today, but water once flowed over the surface of the planet, creating channels. Planetary geologists estimate that the riverbeds have been dry for more than a billion years.

◆ Top right: There still is water on Mars, but it is frozen in the polar ice-caps and as permafrost underground.

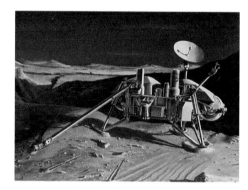

which successfully landed there in the summer of 1976. They were the first functioning devices from Earth to land on Mars.

The Vikings photographed the desert landscape, analyzed the air and took temperature, air-pressure and wind-velocity readings. The results confirmed previous knowledge of the thin, dry atmosphere, which is almost pure carbon dioxide. To no one's surprise, the photographs showed no evidence of life.

Next, small soil samples were examined in four experiments designed to test for life as we know it. No recognizable life forms were found. The Viking probes revealed that Mars is far less friendly to life than the bleakest Antarctic ice fields.

However, there are good reasons to think that life may have emerged on Mars about the same time it began on Earth. Between three and four billion years ago, Martian volcanoes probably spewed enough water vapour over the planet to create an ocean more than 30 metres deep. That's when the riverbeds must have been carved out. Now, though, what water is left is frozen in the polar caps and as permafrost beneath the surface.

Did life forms develop before all the water froze? Until Mars is explored more thoroughly, we will have to wait for the answer to that question. But if future explorers do discover evidence of past life on Mars, it will tell us that life as we know it is not limited to the surface of Earth.

◆ Bottom left: Photograph of Mars taken by the Hubble Space Telescope shows the planet's polar cap and its yellow dust-laden deserts. The dark areas are less dusty rocky regions.
◆ Bottom right: The surface of Mars photographed by a Viking space probe in 1976. Parts of the spacecraft are seen in the foreground.
◆ Top left: This NASA illustration of a Viking space probe on Mars depicts the robotic arm used to collect soil samples. The painting, which shows a blue sky, was made before the Viking landers arrived at Mars. They revealed that the Martian sky is a peach colour, the result of windblown dust.
◆ Top right: Future explorers on Mars search for signs that life once existed on the planet. Fossil evidence appears in the foreground.

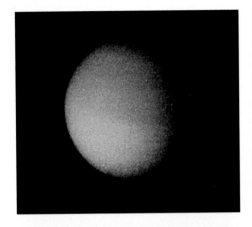

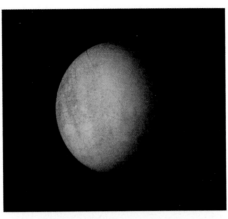

One of the most successful space missions of all time was the odyssey of Voyager 1 and Voyager 2. The two American robot space probes flew past Jupiter in 1979 and Saturn in 1981 and 1982. Voyager 2 went on to explore Uranus in 1986 and Neptune in 1989. These journeys revolutionized our knowledge of the solar system's giant planets and the dozens of moons that orbit them.

The moons of the giant planets had never been seen in detail before, and the cameras aboard the two Voyagers revealed landscapes that nobody had imagined. Among

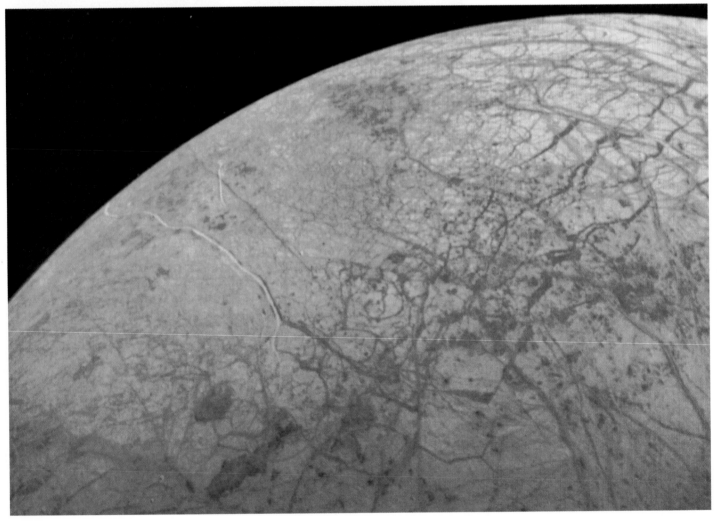

◆ Top left: Saturn's largest moon, Titan, is the only moon in the solar system that has a substantial atmosphere. The atmosphere contains a thick layer of mist and clouds that blankets the entire world.

◆ Top right: Jupiter's moon Europa is the same size as the Earth's moon.
◆ Bottom: Europa's surface looks like a cracked eggshell. What we are seeing is actually a crust of ice, perhaps 50 kilometres thick, that may cover a global ocean of water. (All the photographs on this page were taken by the American Voyager 2 space probe.)

the biggest surprises were Jupiter's moon Europa and Saturn's moon Titan. Europa is about the same size as the Earth's moon, and Titan is a bit larger. The similarity ends there, however. The moons of the outer planets are composed mainly of ice, rather than rock, but Europa and Titan are not crater-covered ice balls like most of their companions.

Europa is the only body in our solar system, apart from Earth, that has large amounts of liquid water. But unlike the water on Earth, the liquid water on this large moon is not on the surface. It is so cold out at Jupiter's vast distance from the sun that Europa's surface temperature is minus 120 degrees C, and its oceans lie beneath an icy mantle.

The ice forms a crust around Europa perhaps 50 kilometres thick. The global ocean below could be up to 10 times deeper than the deepest oceans on Earth. Beneath the ocean, a core of rock is heated by decaying radioactive rocks and by the forces of tides produced by the gravitational pull of Jupiter's other large moons.

Europa's warm core keeps its ocean liquid. Since liquid water is the key ingredient for life as *we* know it, our question is, Did life ever develop in Europa's ocean? We don't get many clues from looking at the surface of Europa, which resembles the shell of a hard-boiled egg scored with hairline cracks. The icy skin has buckled as a result of the forces from below and from the occasional comet crashing into it.

Titan is a totally different world, at least on the outside. Spacecraft photographs show that it is completely cloaked in mist, smog and clouds. It is the only moon in the solar system with a substantial atmosphere and the only known celestial object whose atmosphere has approximately the same density and surface pressure as the Earth's. Titan's icy surface may have lakes of liquid methane, and methane geysers may fountain up from within.

The Earth's atmosphere is 78 percent nitrogen and 21 percent oxygen, while Titan's is about 95 percent nitrogen and 5 percent methane, which we know as natural gas. The surface atmospheric pressure on Titan is 1½ times that of our planet. The big difference is temperature. The Earth's average surface temperature is just above the freezing point of water (0 degrees C), while Titan's, at minus 180 degrees C, is barely above the freezing point of methane.

As water on Earth exists in solid, liquid and gaseous states, so methane on Titan is solid, liquid and gas. There must be glaciers of methane ice, lakes or ponds of liquid methane and methane vapour in the lower levels of the nitrogen atmosphere that could fall as methane rain. Methane serves the same role on Titan as water does on Earth, and it is even possible that a methane-based form of life could exist.

Humans protected by well-insulated heated spacesuits would have no problem exploring Titan. By the late 21st century, astronauts from Earth will walk on the surface of this intriguing world.

◆ The surface of Titan will probably be explored by humans sometime in the 21st century. Here, two intrepid astronauts cautiously approach a geyser spewing liquid and vaporous methane from an underground pool.

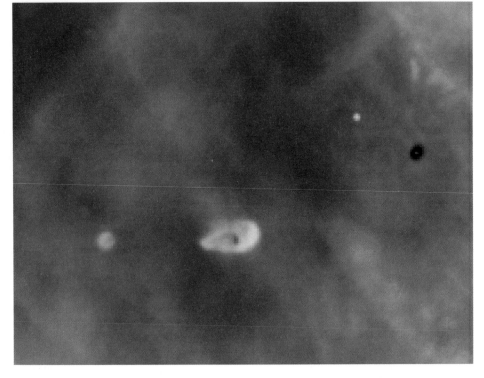

About 4½ billion years ago, a giant cloud of gas and dust in one of the spiral arms of the Milky Way Galaxy began to collapse. The cloud was enormous, containing enough material to make thousands of stars.

In one pocket of the cloud, material spun itself into a thin disc like a miniature spiral galaxy. Matter at the centre of this disc formed into the sun, and farther out, it collected into innumerable small objects, called planetesimals, ranging from peanut-sized pieces to bodies hundreds of kilometres in diameter. As legions of planetesimals orbited the new sun, they gently collided with each other, often sticking together to form larger objects. Eventually, from this material, planets were born.

The formation of Earth probably took less than 40 million years. Just around the time it was completed, a smaller planet about the size of Mars smashed into Earth. Huge amounts of molten material from the collision splashed into the space surrounding Earth. The debris soon formed a ring around our planet, and material from that ring collected into the primal Earth's satellite—the moon.

For about 600 million years after their formation, Earth and moon were steadily bombarded by the remaining planetesimals. The moon's cratered face dates from this era. Finally, around 3.6 billion years ago, things settled down enough for life to develop in the Earth's oceans.

Geologists have measured the ages of moon rocks and meteorites (planetesimals from the asteroid belt between Mars and Jupiter that have plunged through the atmosphere to the Earth's surface). Nothing is older than 4.6 billion years, which strongly suggests that all objects in our solar system were formed around that time.

The idea that the planets emerged from a thin disc of material swirling around the sun is supported by the fact that all the planets orbit the sun in the same flat plane, like balls rolling around on a billiard

◆ Top: The Orion Nebula is a stellar nursery, a giant cloud of gas and dust where new stars are constantly being born. Astronomers suspect that our sun and its family of planets emerged from a nebula like this almost five bil-lion years ago. The nebula is visible because it is being illuminated by stars born during the past few million years.
◆ Bottom: This high-power view of a tiny section of the Orion Nebula was taken by the Hubble Space Telescope. It shows two small bright patches and one dark one that scientists believe are solar systems in the process of forming. Each patch appears to be a new star surrounded by material that will eventually become planets.

table. It is difficult for us to imagine how the planets could have become this organized unless they were born that way.

Could planets form around other stars? The Hubble Space Telescope has provided some of the most convincing evidence that planets probably do form around many stars.

Astronomers have known for years that the Orion Nebula and similar clouds of celestial gas and dust are stellar nurseries in which thousands of stars have been born. In 1994, Hubble's powerful optics zoomed in on a small section of the Orion Nebula and revealed never-before-seen details around newborn stars.

About half of the stars examined had cloudlike discs or doughnuts around them, which is exactly what we think our sun looked like during the period when planets were forming from the disc of material surrounding it. This is persuasive evidence that the same thing is occurring around these new stars as happened more than four billion years ago during the formation of our own solar system.

In 1991, two planets somewhat larger than Earth were discovered orbiting a neutron star. The discovery caused a sensation. By timing variations in the pulses caused by the rotation of the neutron star, astronomers were able to calculate that the star was being affected by the gravity of two planets, 3.4 and 2.8 times the Earth's mass. They have orbits similar to the orbits of Mercury and Venus in our own solar system.

However, a neutron star is the dense core of an exploded star (see page 55), and its planets probably formed from the debris left over after the explosion. Astronomers consider these bodies to be interesting but completely unrelated to planets like those in our solar system. So far, no planets have been found orbiting a normal star like the sun, even though we suspect they must be there. The search for such planets is one of the top priorities in astronomy today.

◆ Top: When our sun was born, material remaining around it swirled into a disk of planetesimals, which gradually collected to form the planets, including Earth.

◆ Bottom: This illustration shows Earth not long after the formation of the moon. Leftover material from the creation of the planets is seen bombarding Earth and its moon. During this period, the Earth's surface was like a sea of lava. It took millions of years before a solid rocky mantle formed. Later, water emitted as vapour from volcanoes began to fall as rain to create oceans. Other water was added when water-ice comets crashed into our planet.

What type of planet will extraterrestrial life inhabit? The complete range of possible habitable worlds is depicted in the illustration on the facing page.

The three largest planets represent gas giants, similar to Jupiter, Saturn, Uranus and Neptune in our own solar system. Gas giants are essentially huge balls of varying proportions of hydrogen, helium, methane and ammonia gases and water-ice slush. They form at great distances from their parent star and are little affected by any erratic behaviour by that star. Unlike the situation with terrestrial planets (see below), a gas giant in one solar system is likely to be very similar to a gas giant in another solar system.

The fourth largest world, a transitional celestial body between terrestrial planets and gas giants, does not exist in our solar system. But there is no reason why it couldn't exist elsewhere. It would have a dense atmosphere and probably a liquid-water surface, which, regardless of its distance from the sun, would be kept liquid by the planet's hot rocky core. It would be a prospect for both ocean and atmospheric life.

The next three worlds are Earthlike terrestrial planets that we consider to be the most favourable habitats for intelligent extraterrestrials. These worlds have four crucial ingredients: (a) an optimum location around the parent star to retain abundant liquid water on the surface; (b) a substantial atmosphere; (c) a significant land surface for life to advance onto from the oceans; and (d) gravity ranging anywhere from one-half to three times the gravity of Earth. Perhaps one star in a thousand has such a planet. Maybe one in a million. No one knows—at least, no one on Earth.

Continuing down the chain, planet number eight is a "wet Mars." Mars may have been like this three billion years ago, before its atmosphere thinned and its surface water permanently froze underground.

The next six objects in the sequence have been termed *iceteroids* by coauthor and illustrator Adolf Schaller. They are primarily water-ice bodies, similar to the moons of the giant planets of our solar system. Depending on the local conditions, iceteroids may sustain liquid-water environments below the surface. Jupiter's moon Europa (see page 20) is a prime example of this group. Rocky bodies in the same size range, such as the asteroids of our solar system, are not included in this illustration because, having neither atmospheres nor water in any form, they are unsuitable environments for life as we know it.

The final five objects are large comets. Comets are chiefly water ice, but heat from the sun can melt portions of them. Molecules that form the basic building blocks of life (combinations of hydrogen, carbon, nitrogen and oxygen atoms) have been discovered on comets. One intriguing theory is that Earth was "seeded" with these molecules during the heavy bombardment of comets four billion years ago.

Worlds with many other surface conditions exist, but these are the ones on which we would be most likely to encounter extraterrestrial life.

The Terrestrial Planets

The four inner planets of our solar system—Mercury, Venus, Earth and Mars—are also known as terrestrial planets, because they are composed primarily of rock and metal. None of the Earth's companions, however, is very Earthlike in terms of surface conditions. (The Earth's moon is shown here as well.) Venus, seen just to the left of Earth in this NASA mosaic, is the only planet that closely matches Earth in size.

Astronomers think Venus had oceans early in its history, but because it is only two-thirds the Earth's distance from the sun, solar heating long ago boiled away the water, leaving a thick, searing-hot atmosphere. Venus dramatically demonstrates that an Earth-sized planet is not necessarily an Earthlike planet. Mercury, shown at far left, is a cratered world similar to the moon. Mars, at lower right, is described on page 18.

◆ All types and sizes of planets that might possibly support life as we know it are represented here. They range from huge gas giants like Jupiter, Saturn and Neptune to comets the size of large mountains. In the middle of this range are the Earthlike worlds. Possible life forms that could have evolved on the planets shown here are explored on the following pages. Incidentally, no matter how massive a gas giant planet might be, it would not be much larger than Jupiter. As mass is added to a Jupiter-type planet, gravity increases, which tends to compress its gases. The result is more matter in a globe approximately the same size.

◆ The existence of the moon may
have been crucial to the development
of life on Earth. Without it, our
planet's rotation axis might wobble,
which in turn could cause extreme
fluctuations in the Earth's climate.

Earth is unique among the four rocky terrestrial planets of the solar system because it has a large moon—one-quarter the Earth's diameter. Mars has two moons, but they are barely more than flying mountains. Mercury and Venus have none.

What does having such a huge moon have to do with life? Probably plenty. According to recent research, the existence of the moon has stabilized the Earth's axis tilt, which has allowed our planet to maintain a climate comfortable for life for billions of years.

Apparently, the moon's gravity and its nearly circular orbit around Earth act in the same way as the balance pole carried by a tightrope walker. Without it, the Earth's axis tilt would wobble, producing wild climate changes over cycles as short as a few million years. Higher forms of life might never have evolved on Earth if it were not for that large moon.

At present, the Earth's axis is tipped 23½ degrees from vertical, an angle that produces the annual progression of seasons. The Earth's axis tilt has stayed within a few degrees of this value for hundreds of millions, possibly billions, of years. But without the stabilizing influence of the moon's gravity as it circles Earth each month, the Earth's axis would swing from 0 to 85 degrees in a chaotic manner. A shift of 10 to 20 degrees could occur in just one million years.

If the Earth's axis were tilted by 85 degrees, the planet would be lying on its side. The north and south poles would face almost directly toward the sun during summer in each hemisphere. Over most of the planet in the summer months, the sun would never set. Winter would include total darkness for several months each year.

Climate changes under such conditions would be enormous and devastating. How could evolution proceed beyond primitive life forms on a planet where the climate in any particular place could radically change

from humid to desert or polar to subtropical conditions in a million years—a mere swing of the pendulum on a cosmic time scale?

What makes this aspect of the Earth-moon system so important is that it applies to any Earthlike planet which has a relatively rapid spin rate. Mars, for example, has a daily spin rate just 40 minutes longer than the Earth's, but since it has no large moon to keep the

axis-tilt oscillation in check, its axis tilt varies between 10 and 50 degrees over periods as short as a few million years.

Because Venus and Mercury have extremely slow rotation periods, several Earth months long, they do not wobble like Mars. (But slow rotation periods in themselves doom a planet to baking on one side and freezing on the other during the long daynight cycle.)

All of this would have little meaning if large moons like the Earth's were typical. But recent studies of the origin of the Earthmoon system indicate that our moon's existence is a long-shot accident that began with a collision about 4½ billion years ago between Earth and a Mars-sized body. The collision sprayed huge quantities of material into orbit around Earth, material that would later collect to form a large moon.

These studies imply that life as we know it on an Earthlike planet might be limited to a rather narrow range of circumstances and fortunate accidents. Other Earths, teeming with life, may be rarer than we once thought.

◆ Top: Extraterrestrials approaching Earth and the moon would see them just as they appear in this photograph taken by the Galileo space probe from six million kilometres away. This picture shows, better than any other, the correct size comparison between the two worlds.
◆ Bottom: The moon's rocky, lifeless surface contrasts starkly with the clouds and water of Earth, a planet teeming with life. From what we know today, the vast majority of rocky bodies in the universe will prove to be much more like the moon than the Earth. It takes a rare combination of circumstances to create a planet with liquid water over most of its surface.

Intelligence seems to be a powerful tool for self-preservation. In any particular species, the smart guys get food, while the stupid ones die off. This appears to be a fundamental rule of nature on Earth, and it results in larger and larger brains as life evolves over broad spans of time.

If we discover life on another planet, we might also expect to find that the most successful creatures on that alien world are those with the largest brains and the greatest ability to think their way out of trouble and into the next meal. (Of course, if they are advanced beyond us, they may be using their reasoning powers for pursuits we can scarcely imagine.)

The illustrations here depict two examples of the emergence of intelligence. Below is a chase scene similar to those which have been repeated countless times on Earth. Here, it is played out between two unearthly creatures.

The planet on which this life-and-death struggle takes place is slightly smaller than Earth, with about 80 percent of the Earth's gravity. As on Earth, its animal life is clearly divided into predators and prey. The predator in this case is a terrifying 30-foot-tall (9 m) flesh-eating biped of enormous strength and speed. It is capable of running down and sinking its claws and teeth into any creature on the planet. The other animal, a herbivore, is an impressive beast in its own right, standing 20 feet (6 m) high from its toes to the top of its head, but the docile plant-eater is no match for the great carnivore.

Because of the planet's lower gravity, the formidable toothed carnivore is significantly larger than any similar animal that ever walked on Earth, including *Tyrannosaurus rex*. Lower gravity means less weight, so plants and animals can reach a larger size.

On Earth, herbivores outnumber carnivores by about 10 to 1. This allows the carnivores to pick and choose their meals, and they usually weed out the weak and the sick first. The savage beast depicted here has no natural enemies on this planet. It will meet its doom only by running out of food or from disease or natural disaster, such as climate change or an asteroid impact. Carnivores also keep herbivores from multiplying and overrunning the planet. Such balances of nature seem to operate at all levels on Earth, and there's no reason to expect that this fundamental arrangement does not exist on other worlds that harbour living creatures.

The illustration at right shows an Earthlike planet with slightly less mass and lower gravity than Earth—perhaps the same planet where the great carnivore once ruled, but millions of years later. Having mastered the manipulation of fire, these creatures gather around the hearth. They share a common interest in survival and are beginning to cultivate a sense of sharing and cooperation that may eventually lead to what we regard as a civilization.

The large moon looming in the twilight sky may have been the key to stabilizing this planet's climate over vast spans of time, as explained on page 27. It also provides the most obvious inspiration in the night sky—a mysterious celestial lamp to excite curious minds.

◆ A life-and-death struggle plays itself out on the plains of an alien planet in some far-off corner of the galaxy. Scenes such as this are purely imaginary, but if higher forms of life develop on other worlds, we would expect some basic rules to apply. One of them is that living things need food. Collecting food requires intelligence. The smarter a creature is, the more ways it can devise to secure a meal. Intelligence is, therefore, a basic advantage among living things. So it is not unreasonable to think that we might find examples of highly intelligent creatures elsewhere in the universe.

◆ Have any other creatures in the universe reached a level of intelligence comparable to that of humans? Are there extraterrestrials out there that have long ago surpassed our level of thought? How many civilizations have gazed at the starry night sky and wondered whether they are alone? These are powerful questions that may never be answered.

◆ The lovable E.T., from the film
E.T. The Extra-Terrestrial, in some
basic ways resembles a human baby
—and is as endearing and almost as
helpless as one. Is this a body shape
that makes sense?

One billion years ago, the smartest creatures on Earth were worms with brains so small, they would hardly be visible to the naked eye. Six hundred million years later, fish were the most intelligent life on the planet. Reptiles claimed that honour roughly 300 million years ago, and dinosaurs were the brain champs from about 200 million years ago until they abruptly disappeared from Earth 65 million years ago.

During their time on Earth, the dinosaurs were totally dominant, while mammals were the size of mice. Once the larger-brained

dinosaurs were out of the way, however, mammals took over and steadily advanced to where we are today.

But a large brain alone does not necessarily indicate a high degree of intelligence. Elephant brains are bigger than human brains, which makes sense; it requires a lot of brain capacity to maintain such a huge body. This is not to say that elephants are not clever; it is just a reminder that the crucial measure is brain size compared with body size. A big brain in a moderate-sized body means there is more brain capacity available for thought. When ranked according to a brain-to-body-size ratio, humans are first, dolphins second and chimpanzees and gorillas third.

This ranking tells us several important things. The first is that, on Earth at least, land creatures develop much larger brains than creatures that evolve in water. What about the dolphin? Although it lives in the sea today, the dolphin was once a land animal, something like a wolf or a small bear, which gradually evolved to a watery existence about 50 million years ago. The shark is the smartest creature to evolve in the sea, but by land-animal standards, it is dull-witted.

Life on land is far more demanding than life in the sea. It requires greater ingenuity to capture food, and the climate is more variable than underwater. Devising survival strategies on land has been the driving force behind the evolution of bigger brains on Earth.

For land-dwelling creatures with large brains, the humanoid form has many advantages. For one thing, it frees the upper limbs —the hands—to make tools and develop technology. Also, if the head is anywhere but on a short neck directly atop the body, the bone structure and muscles required to support it are more massive and therefore less versatile than the humanoid's. We think it may be reasonable to assume that any land creature which walks on two legs and has a heavy head because of a big brain would be humanoid in form no matter what it initially

evolved from. As we will talk about on page 32, large-brained humanoid creatures might even have eventually evolved from dinosaurs, had the dinosaurs not been wiped out. (In the popular movie *E.T. The Extra-Terrestrial*, the humanoid creature, E.T., seems to have evolved from a turtlelike reptile.)

Likewise, there is an optimum shape for advanced ocean life. Dolphins and fish have the same basic body shape and are similar in appearance to aquatic dinosaurs that have long since become extinct. The same pattern holds true for flying creatures: bats, birds and pterodactyls. Certain body shapes are ideal for certain functions, but for large-brained land animals, the most efficient shape may be the basic humanoid form.

Babies and E.T.

Fictional intelligent extraterrestrials are often simply outer-space versions of human babies. The big head and large, bright eyes of a human baby are the distinguishing features of E.T., facing page, and of the visitors from the film *Close Encounters of the Third Kind*, page 15. How could they be anything but friendly?

◆ Top: Certain body shapes suit certain environments. On Earth, the most successful creatures that live in water look like fish. Even the dolphin, which was once a land animal, has evolved over the past 50 million years to look like a fish, because that's the most efficient way to get around in water.
◆ Bottom: For carrying around a large brain in a body that has to survive on land, a humanoid form works best— at least on planet Earth.

Suppose dinosaurs had not become extinct? While we can merely guess how extra-terrestrials might look, we have a hint of what intelligent life on Earth might have been like if the history of life on this planet had been changed just slightly.

Many scientists believe that about 65 million years ago, a giant asteroid or comet smashed into Earth, exploding upon impact with the force of millions of hydrogen bombs. The explosion alone would have killed countless life forms. Dust thrown into the atmosphere by the impact caused a drastic change in the Earth's climate that eventually wiped out more than half of the life forms on the planet, including all the dinosaurs. Nothing weighing more than 25 kilograms survived.

But what if this event hadn't happened and the dinosaurs had continued to evolve? That's what Dale Russell, a palaeontologist at the Canadian Museum of Nature, wondered. Russell theoretically extended the evolution of the most intelligent known

◆ The most intelligent life on Earth might have looked something like this if dinosaurs had not become extinct. The creature, called Dinoman, was created by simulating what might have happened if the smartest dinosaurs had continued to evolve for millions of years. It is warm-blooded and about the size of a 13-year-old human. What this research suggests is that the humanoid shape might be a natural form for a creature with a large brain.

Dinoman is what might have been. This is one scientist's idea of what the smartest creature on Earth might have looked like if the dinosaurs had not disappeared from the planet.

33

dinosaur, a long-tailed forest dweller about five feet (152 cm) tall called *Stenonychosaurus*. This dinosaur had the largest brain, compared with body weight, of any animal on Earth.

After forecasting 50 million years of theoretical evolution, Russell came up with Dinoman, a hairless green-skinned creature with a bulging skull, luminous catlike eyes and three-fingered hands, not unlike some of the extraterrestrials that have populated science fiction films.

Dinoman is 4½ feet (137 cm) tall and would have a live weight of about 32 kilograms. Its brain is the same size as that of a human of similar stature. Since the teeth of *Stenonychosaurus* were small compared with related dinosaurs, Russell thinks that teeth may have been on the way out from an evolutionary standpoint. Dinoman, therefore, has none. Instead, the biting edges of the mouth are "keratinous occlusal surfaces," similar to those of a turtle.

Russell contends that there is no biological reason why the evolution of the dinosaurs to Dinoman could not have taken place, if only the dinosaurs had not been wiped out and replaced by the mammals.

Expanding on this idea, we can speculate that evolution leading to a gradual advancement of intelligence may be the normal result of the progress of life anywhere in the universe. Evolution is a natural process by which different species adapt physically to a variety of changing conditions. Scientists believe that living creatures unknowingly "select" new characteristics which will enable them to live in their environments more successfully. Each generation, whether human or animal, reveals its own variations. They may be insignificant—slight changes in hair or eye colour, for example—but others, like intelligence, extra muscle strength and quick reflexes, can prove to be distinct advantages.

Animals that happen to have such advan-

tages are more likely to survive, find strong mates and pass these traits on to their offspring. Thus from generation to generation, the most successful creatures—the smartest, the strongest and the quickest—will increase in number while the less successful will decrease.

Throughout the Earth's history, creatures have become smarter, the inevitable result

of evolution. Likewise, the general body form of humans—two arms, two legs and a head on a relatively short neck—is no accident. It is the most logical arrangement for a big-brained land-dwelling creature.

If evolution is a key chapter in life's strategy for survival anywhere in the universe, then we can learn a great deal from its patterns here on Earth.

◆ *Stenonychosaurus* (sten-NON-ik-o-sore-us), at right in this picture, was about the size of a kangaroo and lived 70 million to 80 million years ago in what is now western Canada. It was the smartest dinosaur known, with a larger brain (compared with body weight) than that of any other animal on Earth. The model of Dinoman is shown beside *Stenonychosaurus*.

Many animals, including humans, rely on five basic sensory systems to gather information about their environment. The parts of the body associated with sight, hearing, taste, smell and touch are sensors that receive information from the outside world and relay it to the brain for analysis.

Extraterrestrials would need sensors too. Perhaps they could get by with fewer than five; perhaps they would require more. The following is a compilation of sensors, some used on Earth and some that extraterrestrials might find useful.

A. Sight

The eye that is most familiar to a human is our own (1), which has a high-resolution lens and a focusing capability for seeing detail either up close or far away.

In the compound eye (2), such as that found on the common housefly, each of hundreds of lenses is at the end of a stalk, and the stalks are gathered together in a bundle. Each lens delivers an image of a small portion of the environment to the creature's brain. A combination of all these small pictures allows for good wide-angle coverage, but the compound eye is poor at resolving specific detail. Its real strength lies in its ability to detect movement.

Even less complex in design is the simple eye (3), which is usually found in clusters or rows. Simple eyes reveal light or dark and a few shades in between but very little detail. Some insects and sea creatures have rows of simple eyes down their backs or on the backs of their heads, giving them crude wide-angle vision.

Spiders use multiple placement of simple lensed eyes (4) with overlapping fields of view to determine the position and the approximate distance of objects.

Eyes with narrow-slitted pupils (5) provide highly focused detail in either a horizontal or a vertical plane, depending on which way the slits are oriented. Eyes designed for

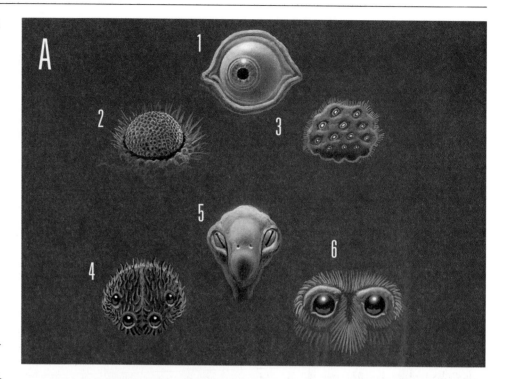

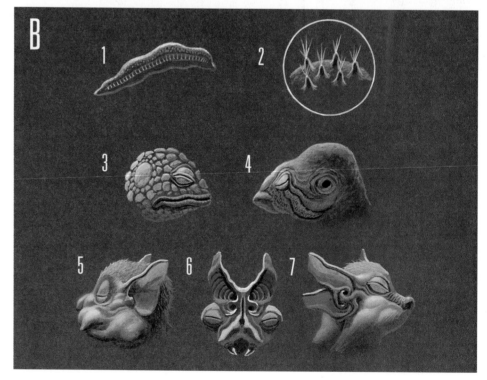

◆ Sight, hearing, taste, smell and touch are the five basic ways by which humans and many other Earth creatures explore their environment. Alien sensory systems would likewise be adapted to the specific conditions of the aliens' planet. Extraterrestrials could be bug-eyed and adorned with antennae, but more likely, they will perceive their environment in ingenious ways that are beyond our imaginations.

night vision (6) have very large pupils for high-resolution dim-light viewing.

Whether on Earth or on the planet of another sun, two eyes seem to be a practical idea. Two eyes provide excellent distance determination, because they produce a three-dimensional image from which the brain can calculate the comparative distance to the object.

B. Hearing

In humans, sound waves are collected by the ear. They then travel down a short canal and strike the tympanic membrane, or eardrum, causing it to vibrate at the same rate as the arriving sound waves. These vibrations are instantly amplified and transmitted to the brain.

Some animals have a hearing membrane on the surface (3) or tiny hairs (2) that vibrate when sound waves pass over them. Others have a simple opening (4) that leads to an inner eardrum. As with the simplest eyes, which distinguish only light or dark, the simplest ears would consist of an array of nerves (1) on the surface that would be sensitive to pressure waves from nearby sounds.

The more advanced creatures on Earth have an ear structure that acts as a dish to focus the sound (5). Specialized ultrasonic (very high-frequency) receiving ears (6) are what a bat uses for aerial sonar. The bat sends out pulses of sound pitched far above human hearing. This sound bounces off objects and is then picked up by the bat's sensitive ears. In this way, a bat can determine with great accuracy not only what is in front of it at night but also how far away it is.

Our final creature (7) has a complex multifrequency-sensitive ear with specialized lobes that collect signals for separate internal ears designed to detect specific sound-frequency ranges.

C. Taste

Taste operates in conjunction with smell. Any loss in our sense of smell, as sometimes

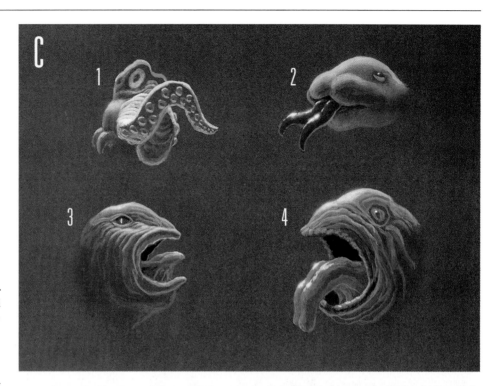

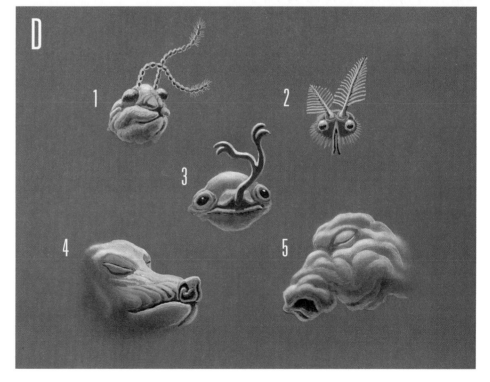

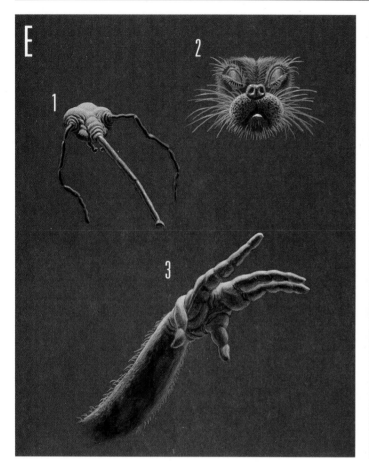

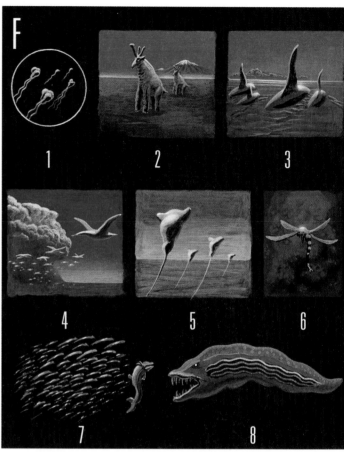

happens during a bad head cold, greatly reduces our sense of taste.

Taste does not necessarily have to occur in the mouth. An elephantlike trunk (1) could contain the taste sensors. Or a tongue extended outside the mouth (2) could gather tiny amounts of chemical substances contained in vapours or mists, which could then be tasted inside the mouth when the tongue is retracted. In this way, a creature could compensate for having an underdeveloped sense of smell. It could even taste a potential meal without touching it.

This creature (3) has a tongue and taste capacity similar to that of humans, while its neighbour (4) possesses a supersensitive taste detector and is able to taste the air outside the mouth if it contains suspended particles or droplets.

D. Smell

Because chemicals and substances from plants and animals are suspended in the atmosphere near ground level, specialized receptors in antennae (1 and 2) and tongues (3) act as "noses" for some insects on Earth. Such a system could be used by extraterrestrial life. A general-purpose nose (4) is compared with a highly sensitive snout (5) designed to identify very subtle smells.

E. Touch

Tentacles and antennae (1) and hairs and whiskers (2) are effective advance-warning systems, especially if the other senses are not particularly acute. In humans and in many other species on Earth, touch is most sensitive through the hands (3), but because the entire exterior of the body has a certain sensitivity, touch is an all-body sense.

F. Magnetic-Field Detection

Although humans have retained the senses of sight, hearing, taste, smell and touch from their evolutionary ancestors, other creatures on Earth have developed a sixth sense: magnetic-field detection. The Earth's magnetic field causes a compass to point to the magnetic poles. In the same way, animals that can sense the magnetic field can use it to determine direction.

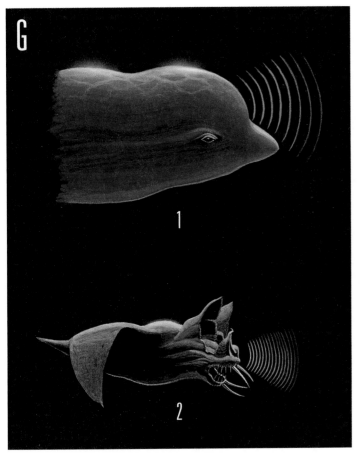

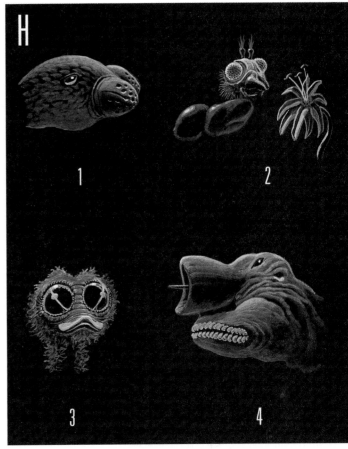

Some insects, fish and birds have sensors in their brains that use the Earth's magnetic field for navigation and orientation, particularly during periods of migration. If an Earth-like planet were to have a stronger magnetic field than the Earth's, magnetic-field detectors would probably be a more prevalent sensory system in that planet's life forms than on Earth, because it would be easier to sense.

Shown here are microorganisms (1), land animals (2), marine animals (3), aerial creatures, both winged (4) and ballooned (5), and insects (6). All these life forms could use magnetic-field detection to find their way from one feeding ground to another. Ocean creatures in a stronger magnetic environment might communicate direction and

move as a coherent mass (7) even more effectively than fish in schools do as a means of self-protection. Finally, we see a vicious-looking water creature (8) that can discharge an electric-current shock to stun its prey.

G. Sonar

Sonar is a contraction of the words "*so*und *na*vigation *r*anging." In nature, it is a substitute for sight to detect the location and distance of objects. On Earth, sonar is used by dolphins (1) and bats (2). See also "Hearing," page 35.

H. Non-Optical Radiation

The human eye is sensitive to what we call visible light. Scientists call it optical radia-

tion, but there is a whole spectrum of other radiation: infrared, ultraviolet, microwave, radiowave, x-ray and gamma ray.

Infrared radiation is heat radiation. Humans can feel it at close range, but creatures with infrared-sensitive eyes (1) could scan the landscape to detect the body warmth given off by other animals. Some alien life forms (2) might specialize in ultraviolet-light imaging, as have some insects on Earth, which enables them to locate certain plants that are bright in ultraviolet light. A microwave-detection vision system (3) would be effective on a planet with a thick, smoggy atmosphere. There might be distinct advantages in an ability to generate and transmit microwaves (4).

Some years ago, a famous astronomer was giving a public lecture at a big-city planetarium. His topic was life on other worlds, and the auditorium was packed. Everyone listened intently as he explained that our galaxy, the Milky Way, contains hundreds of billions of stars and that many of those stars could have planets orbiting them, just as our sun does. After displaying several charts and tables, he concluded that some of the planets of other suns—perhaps as many as a million in our galaxy—could support intelligent life.

During the question period, a member of the audience congratulated the astronomer on his eloquent discussion of the possibility of life on other worlds but then asked: "What I want to know is, Will they be as ugly as my Uncle Ralph?"

Of course, any attempt to show the physical features of extraterrestrials is mere conjecture, but it's fun to try, because the possibilities are infinite and the exercise stretches our minds to encompass many alien environments.

Let's begin with the beaked "feathered" fellow at upper left, facing page. The large eyes suggest that this creature is most active in low light or at night. As on a terrestrial bird, the mouth consists of a tweezerlike appendage, a feature that automatically restricts its ability to grasp objects. Speech would be limited to pitch variations.

Not so the alien at upper right, facing page, which sports a complex mouth structure that provides much finer grasping ability and far more facility in vocal modifications. The mouth is a baglike muscular membrane with touch-sensitive pseudofingers at the top and bottom of what might be called its lips. With these, it has a more developed manipulative ability than the beaked critter has. It captures food with quick engulfing lunges of its membranous mouth, similar to the manner in which a butterfly is trapped in a butterfly net. Bag-mouth would be a virtuoso with vowel-type sounds.

The grotesque creature at lower left, facing page, shows us yet another way of grasping. Using sensitive tentacles that spring from the top of its head, this beast can make very delicate movements. Unlike the tentacles of an octopus, which are lined with simple suckers, these tentacles contain sensors— touch, taste and perhaps smell. The "suckers" on this dry-land creature work not by suction but as individual grasping units. The cups lining the "lips" operate in a similar fashion and can grip an object as tightly as any mouth with teeth. The variety of vocal sounds produced by this alien would be basically similar to our own.

The creature at lower right, facing page, is the closest to human in that it sports fingered hands at the end of arms. It has speech capability and, like the other extraterrestrials here, two eyes. A pair of eyes provides essential distance-judgment capability, a huge advantage over a being with a single-eye design.

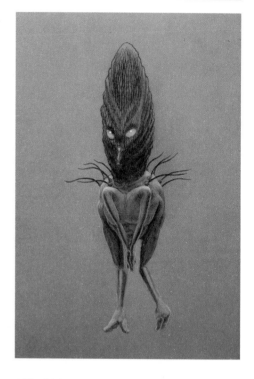

The bizarre cone-head creature above has developed an elaborate signalling mechanism to compensate for weak vocal evolution. It speaks in chirps and whistles but expresses complex emotions and ideas by changing the shape and shading of its head. Flexible accordionlike slats in the head as well as the ability to change colour speak volumes for this fellow. Because manoeuvring with such a bulbous cranium could be awkward, and therefore expose this creature to some danger, it has evolved feelers on its shoulders that can detect anything approaching from behind.

The graceful-looking alien at left lives on a low-gravity world. Its weak muscular development is fully adequate for a planet on which everything weighs half of what it would on Earth.

All these extraterrestrials have manipulative abilities—the use of hands, mouth or other body parts—which is evolution's solution for getting food into the body.

◆ The six creatures illustrated on these two pages represent the diverse forms that intelligent life might take on different planets. The evolution of life on Earth has produced a vast collection of strange and beautiful animals. But that variety might pale in comparison to life forms which inhabit planets we know nothing about—yet.

Everybody has relatives. Extraterrestrials would have them too. Here are portraits from family albums from six different alien planets. No doubt an Earthling would look pretty weird to most of these creatures.

39

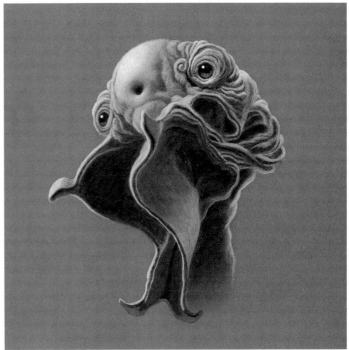

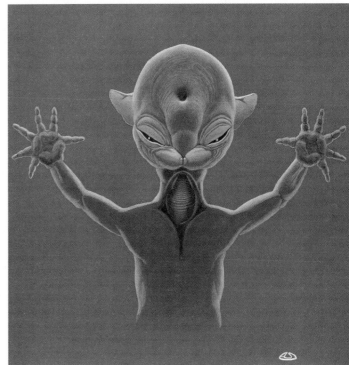

◆ A planet with just half the Earth's surface gravity could have plant and animal life as rich and diverse as the flora and fauna on our planet. But in order for the water to remain liquid, the planet has to have been born the right distance from its parent star. This world has what astronomers call a G8 star, which gives off only half the brightness and warmth of our sun. But the planet orbits around it at two-thirds the Earth's distance from the sun—just the right distance to produce Earthlike conditions. The lower gravity, however, is a contributing factor in the evolution of some very unearthly creatures.

What would it be like to explore a world with only one-half the Earth's surface gravity? Take a close look at the scene depicted on the facing page.

The deep blue sky indicates a thinner, less humid atmosphere only two-thirds the density of the Earth's atmosphere at sea level. This rarefied atmosphere is a less efficient insulation blanket than the Earth's air, so the planet is cooler on average than Earth, even though it orbits a star like our sun at the same distance that Earth does. Oceans cover less than half of this planet's surface.

In this scene, a wide coastal plain has been eroded by wind and water to expose ancient rocks in the cliffs at left. A stream meanders out of the distance from its source in the snow-laden peaks. The tallest trees here tower some 50 metres above the planet's surface. Nibbling on the lower branches are long-necked, two-legged creatures. Superficially, they look something like giraffes, but they are far bigger. Standing 25 metres tall, they could easily peek in the windows of an eight-storey building.

No creature of this height has ever walked on Earth. The trees' natural defence against these herbivores is their height, which keeps their foliage out of reach. The tall, lean bipeds have trunklike probes that allow them to stretch just a bit higher to grasp tender leaves from the giant trees.

The curious creatures on the bluffs in the foreground are cliff crabs, each spanning about a metre from leg tip to leg tip. They deploy sticky silklike strands to snare small flying creatures, such as those seen hovering in dense columns just in front of the more distant bluff. Perched on the cliff, the crabs take advantage of the swarming masses by fishing with their silk lines. A few of them have already reeled in a meal. The intense coloration of the cliff crabs is a warning to potential predators that they are poisonous.

The top surface of the rocky cliff is carpeted with a thick mat of plants, which provides suitable nesting sites for the cliff crabs during the chilly nights. It is morning in this scene, and the land has come alive, rapidly thawing from the near-freezing night temperatures.

A world with a thin atmosphere always has a greater variation in temperature from day to night than a planet with denser air. (An atmosphere acts as a blanket, retaining heat that would otherwise escape into space. The thinner the atmosphere, the cooler the nights.) The plant life has adapted to withstand both the plunges in temperature each night and the relatively dry air, much as in the high deserts of Earth.

A moon, visible beyond the weather-beaten branch at upper left, performs an important function by raising tides to keep the shorelines sloshing. This creates a wider variety of habitats for life. Small ponds, called tidepools, are formed as tides move in and out.

In the far distant past, tidepools on Earth may have been the perfect environment for the formation of complex molecules and the first primitive living matter. They might have played the same role on this alien world too.

The orbiting moon also keeps the interior of this planet in a molten state. A molten interior produces regular volcanic activity, which builds mountains, varies the climate and adds water vapour to the atmosphere in the form of steam. All these factors have helped life become abundant on Earth. We expect they would produce similar results on other Earthlike worlds.

Specifically, this low-gravity planet has a diameter 0.77 times the Earth's diameter and a surface gravity 0.54 times that of Earth. It orbits its sun (a star half as bright as our sun) at 0.62 times the Earth's distance from our sun.

A Smaller Earth

For a planet to be classed as Earthlike, certain basic conditions must be met. First, the planet must have liquid water on its surface—water that stays liquid over billions of years. This almost certainly means that the planet would be orbiting a star with a steady, reliable light output, like our sun. Sunlike stars are called main-sequence stars. There are billions of them in galaxies like the Milky Way.

Second, the planet should have a rotation rate under one Earth week to keep the day/night temperature range from becoming too extreme.

Third, the planet must have a surface gravity that is at least one-half the surface gravity of Earth. Less than that, and the world would probably be more like Mars than Earth. The lower the surface gravity, the easier it is for lighter atoms and molecules in a planet's atmosphere to escape into space. In the case of Mars, the atmosphere is almost gone.

A fourth factor might be the requirement of a large moon (page 27), although this is less certain than the other factors.

The scene on the facing page depicts the morning sun rising in the hazy sky of a planet that has three times the surface gravity of Earth. The atmosphere here is much denser than the Earth's, but like our air, it is rich in oxygen and nitrogen. Much of this planet is covered with water, and lagoons such as the one shown here are common. Most of the creatures on this world live in or near the water.

On this particular day, the lagoon is populated by a herd of large plant-eating animals, something like hippopotamuses, except that they almost never venture onto land. The high gravity makes walking difficult even for the land creatures.

On the shoreline in the foreground at lower right, a pair of squat two-legged creatures are interrupted during a morning drink by the menacing appearance of an aquatic predator—the crocodile of this world—which has the nasty habit of lunging from the water to snatch any unwary animals that

A Larger Earth

In our solar system, there is no planet intermediate in mass between the mass of Earth and that of Uranus, which is 15 times the Earth's mass. But there is no reason why such planets cannot exist. A rocky planet 1.2 times the Earth's diameter (similar to the boxed planet) would have twice the Earth's mass and three times its surface gravity.

come to drink. One of these three-foot-tall (90 cm) bipeds is seen close up at upper right on this page. It is in a resting position, its thick tail supporting its weight of 160 kilograms—triple what the same beast would weigh on Earth.

At upper left is a muscular humanoid just over two feet (60 cm) in height. This stocky fellow is the most advanced form of life on the planet.

In the main scene, at lower left on facing page, a couple of armoured plant-eaters trudge along on six legs. On the opposite side of the lagoon, more bipeds forage on the abundant plant life. The vegetation here has evolved to a fairly sophisticated level. Attractive blossoms promote cross-pollination by flightless insectlike creatures. Very few flying creatures inhabit this world. It is difficult to stay aloft when triple the Earth's gravity is pulling you down.

The trees and stemmed plants are supported by stout, thick trunks, a feature dictated by the high gravity. Things tend to stay short on a high-gravity planet. The bulbous plants in the foreground are not much taller than a fire hydrant. The tallest trees, seen on the horizon, rise no more

than the height of a two-storey building.

The land is well eroded, and although the area we are viewing is fairly wet in comparison with Earth, it is relatively dry by this planet's standards. The whole planet is soggy, humid and dank, with extensive swamps, frequent rainstorms and no mountains like those on Earth.

This planet orbits a star like our sun but at 1.6 times the distance that Earth is from the sun. That would put it slightly beyond the orbit of Mars. Still, the average surface temperature here is higher than on Earth, because the dense atmosphere creates a more efficient greenhouse effect.

The greenhouse effect is produced when certain atmospheric gases (particularly carbon dioxide and methane) act as a blanket, trapping some of the sun's warmth. On Earth, the greenhouse effect adds 30 degrees (Celsius) to the average daily temperature. On our hypothetical triple-gravity world, the atmospheric greenhouse raises the temperature by nearly 100 degrees. But if this is compensated for by the planet's being farther from the sun, the result is a balance that prevents a so-called runaway greenhouse effect, as exists on Venus.

◆This incredibly muscular humanoid, top left, needs all its strength to survive on a planet with a surface gravity three times greater than Earth's. The humanoid coexists on the planet with the stocky biped, top right.

◆ An early-morning scene on an average day on the surface of a world somewhat larger than Earth, with three times the Earth's surface gravity. The rich variety of life here is well adapted to the high gravity, which humans would find crushing. An Earthlike planet with a higher surface gravity than this would likely have much more water and therefore no land above sea level upon which to sustain terrestrial life.

◆ Startled by an intruder, a 30-tonne denizen of an ice-covered Earthlike world bellows forcefully to protect his dominion over a rare opening in the ice. Had Earth been slightly farther from the sun or the sun a bit dimmer than it is, our planet might have experienced a similar fate and been trapped in a permanent global ice age, as is this frigid world.

During the last Ice Age, 18,000 years ago, more than half of North America was covered with ice. In some places, the icy mantle was over three kilometres thick. The average daily temperature was 5 to 10 degrees (Celsius) colder than it is today. Since then, the ice has retreated and the temperature has risen—we are now in an "interglacial period."

Ice ages have swept over Earth many times, but our planet has been able to maintain a delicate balance. Earth has never become a broiling greenhouse desert like Venus. Nor have the glaciers of any previous ice age ever advanced to encase the entire planet. But suppose such a fate were to occur.

Imagine that over thousands of years, the glaciers simply got bigger, pushing ever farther toward the equator until most of the globe was enshrouded in ice. Since ice and snow are highly reflective, the warming rays of the sun would be bounced back into space. The result might be a permanent ice age.

That's the situation pictured on the facing page. Most of the life that formerly inhabited this hypothetical planet has disappeared. Virtually no plants grow in the wickedly cold conditions that have frozen over nearly all the oceans on this world and turned much of it into a global deep-freeze.

But at least one creature has adapted to the frigid environment: a giant ice-eater, 10 metres long and weighing up to 30 tonnes. This gargantuan beast lives mostly underwater and feeds on the undersurface of the ice. There, the ice is encrusted with small organisms that are a rich source of nutrients. Colonies of organisms are most abundant around openings in the ice, called polynyas. Tidal action caused by the large, close moon regularly cracks open the icy crust and creates many such openings.

The ice-eater is equipped with a short, massive, spadelike horn that it uses to chip and pry into the underside of the ice sheet as it feeds. When a chunk of ice comes loose, the ice-eater chews it into manageable pieces for swallowing. The horn is also a powerful weapon in the battle for dominance over rivals or for control of a polynya. The scarring visible on the front of the ice-eater in the foreground is evidence of previous contests. Here, the beast postures as it faces us and bellows a warning, producing a substantial vapour cloud in the process.

Could the fate of planetary glaciation befall Earth? Our planet has managed to remain a suitable habitat for life for nearly four billion years, with liquid water covering much of its surface. If the water hasn't frozen over in that length of time, it likely never will. Even if the oceans on an Earthlike planet eventually froze, some forms of sea life could survive indefinitely, feeding on microscopic organisms that thrive around hot-water vents at the ocean bottom.

The probable fate of planet Earth is that of World F, in the illustration below. Astronomers know that the sun is slowly getting hotter, but Earth still has a billion years or more in its present state.

A Delicate Balance

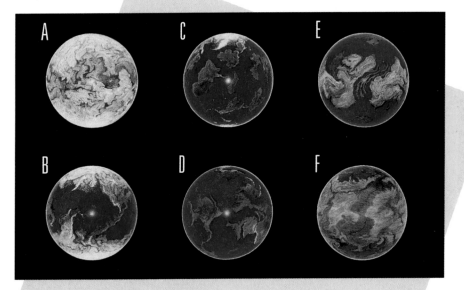

A planet the same size and mass as Earth does not necessarily have similar surface conditions. Venus, the Earth's sister planet, is a perfect example of how different a twin of Earth can be (see page 24). The illustration above presents six hypothetical planets closely resembling but not quite identical to Earth. (For clarity, clouds have been omitted.)

World A is the glaciated world illustrated on the facing page. Only small zones near the equator remain unfrozen. At the peak of the most extensive ice ages on Earth, our planet resembled World B. World C is close to present Earth conditions. During the Age of Dinosaurs, there were no polar icecaps and our planet may have resembled World D. The hot equatorial deserts of World E still leave comfortable conditions near the poles for many forms of life. However, World F experiences temperatures near the boiling point of water at its equator and is approaching the limit as a world on which life could survive. As vast amounts of surface water are boiled off, the atmosphere of World F, laden with water vapour, could start a runaway greenhouse effect, as happened on Venus sometime between three billion and four billion years ago.

A planet similar to Earth could easily have such vast quantities of water on its surface that very little land would poke above sea level. The oceans of this watery world could harbour a wealth of life forms, even though the scarce land areas would remain barren.

The illustration at right depicts a rich community of alien organisms found along relatively shallow coastal margins or island systems similar to the tropical latitudes of Earth. Many of these creatures seem surprisingly familiar, because streamlined forms are the ultimate design for aquatic creatures, no matter where they occur in the universe.

Creatures that move most efficiently through a liquid have a torpedo or bulletlike shape. Other shapes may exist, but the best swimmers will look like fish.

The dolphinlike animals swimming toward us out of the distance can fly in the dense atmosphere of this planet as efficiently as they swim through the water. They are the geese of the sea. Underwater, their powerful dual-action tails, designed for stability when flying, swiftly propel them in the pursuit of food.

Another creature, looking very much like a manta ray, flies effortlessly through the water using huge winglike fins. Here, it is the unfortunate target of a large marine predator,

which displays some of the characteristic features of sharks, orcas and sea lions.

Schools of fishlike organisms dart from the scene of conflict in a closely coordinated mass motion that tends to confuse potential predators. Invertebrate-type animals also take action to remove themselves from danger. Gulping seawater through intake valves, they forcefully expel it through a narrow syphon, using powerful muscular contractions, and jet out of harm's way. Tentacled much as terrestrial squids are, they rapidly retreat by swimming backward.

Finally, two finned crustaceanlike forms are seen in the foreground at bottom foraging for the abundant small organisms stationed throughout the reef. The entire scene is reminiscent of the reef waters familiar to earthly skin divers.

There is virtually no limit to the size that a deep-sea creature could attain. The buoyancy of water makes fish and other sea creatures almost weightless, which, incidentally, is why the space shuttle astronauts train for spacewalks in giant water tanks. Water simulates the feeling of weightlessness that astronauts experience as they orbit in the shuttle.

The creature illustrated on this page, although superficially similar to a whale, is not an air-breather. It is a denizen of the deep, patrolling far below the surface of the ocean world. Dwarfing any terrestrial whale, this hulking beast, hundreds of metres long, has enormous flippers that fold forward to form a gigantic scoop, or funnel. It feeds by passing water through a membranous screen in its body, capturing tiny luminous creatures similar to earthly bioluminescent plankton, which create clouds of diffuse light that illuminate this otherworldly scene in the gloomy depths. The plankton obtain their sustenance from minerals that filter up from hot vents on the seafloor and from a continuous rain of dead creatures and their waste products from above.

◆ A huge extraterrestrial creature of the deep, larger than any earthly whale, ploughs through clouds of planktonlike life forms, gathering a meal as it moves.

◆ There are reasons to believe that a scene depicting life in an ocean on another world, as shown here, would bear many similarities to an ocean scene on our own planet.

In order for vegetation to grow on a planet, liquid water is a crucial ingredient. Without it, life as we know it cannot exist. Because abundant supplies of water have been on the Earth's surface for billions of years, however, vegetation covers vast areas of our planet.

Now, imagine other Earthlike worlds, each warmed by its own sun and likewise nourished by rain, rivers and lakes. These, too, would be thick with alien versions of trees, vines, ferns and undergrowth. Imagine further that you are the first explorer to set foot on one of these worlds.

Your first move is not to move. You stand and look and listen. The air is filled with strange screeches and hoots. Your presence has been noted, but you see nothing. The vegetation is so thick, you can barely take a step. As you contemplate a plan of action, you see a tree branch curl, then uncurl.

It's not a branch at all, but a tail attached to a sinuous monkeylike creature, at upper right, facing page, that has been staring at you, camouflaged in a tree. With arms two metres long—four of them—it is not exactly cuddly. You decide to retreat, only to encounter another tree dweller dangling from a limb, at lower left, facing page.

As your explorations continue, you discover the source of the screeching: a skunk-sized fellow, at upper left, facing page, with a big mouth. But the most interesting creature, at lower right, facing page, and apparently the most advanced in terms of intelligence, is cautious of any approach. It quickly draws back, yet it is too curious to go away.

What would happen if humans could, in fact, explore a world so similar to our own? Would we choose to leave everything as it is for fear of contaminating or harming a planet where millions of years of evolution have been at work?

Interesting question. Extraterrestrials with the ability to visit Earth would face just such a dilemma.

◆ Peeking through the dense forest growth, several creatures of this vegetation-dominated planet prepare to flee if danger threatens.

◆ Inhabitants of a well-vegetated planet range in intelligence from the level of a squirrel, top left, to that of a chimpanzee, bottom right. Life in forests and jungles favours creatures that possess sharp vision and good distance-estimating skills, no matter where in the universe the forests and jungles happen to be.

◆ Peering at us inquisitively, a hulking aerial whale floats in the upper atmosphere of a giant planet that is composed almost entirely of gases. Behind the serene-looking beast are two cities that also serve as launch platforms from which these creatures can escape their planet's atmosphere to explore the realm beyond. Each sunlit dot in the distance is an aerial whale like The Guide.

The solar system's giant planets, such as Jupiter, above, have immense atmospheric cloaks. In fact, these worlds can be regarded as being almost all atmosphere. Deep inside, there may be a small metallic core, but the atmosphere surrounding it is thick, hot and impenetrable. If life as we know it exists on this type of planet, it will be in the upper atmosphere, where room-temperature conditions prevail.

Giant planets are probably common wherever planets have formed around other stars. All the basic ingredients for life (hydrogen, oxygen, nitrogen, carbon and water) occur naturally in the atmosphere of these gas giants. Simple plants might float on the air currents, as seen in the illustration at upper right. Or the life forms might be far more intriguing, as shown on the facing page.

The compelling hulk that greets us is an aerial whale. It appears to be highly intelligent and seems to have come to welcome us personally upon our arrival. We will call this creature "The Guide."

The Guide is more than 500 metres long, a gigantic creature the size of a football stadium whose true dimensions are diminished by the vast open spaces in this scene. Its eyes are at the lower end of the slits; the ears are the slits themselves. The orifices that appear where the ears should be are actually the vocal boxes, which are capable of emitting a deafening roar.

The Guide's mouth also acts as an enormous intake duct to collect air used for jet propulsion. For rapid acceleration, the air is expelled through an opening tucked out of sight underneath. An aerial whale is neutrally buoyant, which means that when it is at rest, it floats like a balloon. It feeds on small airborne life forms.

The giant aerial whales are one of two intelligent life forms on this hypothetical planet. The other consists of relatively tiny crablike creatures that live *inside* the huge beast. The two intelligent beings have managed to develop a shared consciousness that allows them to work together. In this way, they have erected colossal floating launch platforms—spaceports—using a technology that can process membranous plants and other materials found in the atmosphere far below. Since the whale has no arms or manipulative ability, it depends on the crab creatures to build and repair the floating platforms as well as to perform various housekeeping duties within its body. In exchange, the crab creatures are provided with free transportation.

The floating launch complexes (the two light-bulb-shaped objects) are huge cities, some 40 kilometres tall. When launching a spacecraft, a floating city bobs high into the atmosphere so that its top penetrates the stratosphere, from which the launch can take place more efficiently. Meanwhile, the lower bulb still remains comfortably immersed in denser, warmer layers of the atmosphere, where it is easily accessible to commuter traffic.

◆ Left: The giant Jupiter, the largest planet in our solar system, has a thick atmosphere that is thousands of kilometres deep. The planet itself is made up almost entirely of gas, primarily hydrogen and helium. Jupiter is 1,000 times larger than Earth by volume and 318 times more massive. Its colourful visible surface is a thick blanket of frigid clouds, but temperatures deeper in the atmosphere are comfortable by earthly standards. All gas giants have the same basic structure.
◆ Right: Life in the atmosphere of a gas giant planet would be similar in many ways to life in the ocean. Here, on a hypothetical planet, a "flock" of creatures floats in the atmospheric currents.

Living things on Earth are composed almost entirely of a handful of elements: hydrogen, nitrogen, oxygen, phosphorus, sulphur and, most important, carbon. Carbon is the key to bonding the other elements together into giant molecules such as proteins and nucleic acids, which are the building blocks of life as we know it. No other element has carbon's friendly nature, which allows it to link up easily with itself and with many other elements.

While carbon is the vital element for life, water is the crucial environment. Without liquid water, life as we know it perishes. But that's life as *we* know it. What about life as we *don't* know it?

The closest element to carbon, in terms of its ability to bind with other elements into complex molecules, is silicon. Silicon is most commonly found on Earth as silicon dioxide, or quartz, which is the main ingredient in glass. Imagine a planet where silicon plays the role of carbon. If silicon-based life forms could exist, they would more closely resemble crystal than they would the plants and animals on Earth.

That's the scene on this page. The creature that looks like a robotic metal insect is a "lithovore," a hypothetical silicon-based life form ponderously trudging across the crystalline highlands of a world where humans would instantly perish in the inhospitable environment. The lithovore lives in a thick, dense atmosphere in which the air pressure is greater than the water pressure at the bottom of the Earth's oceans. And it's hot here.

Heat comes from deep within the planet, producing temperatures well above the boiling point of water. Fed by "lava" from below, the crystalline landscape is constantly growing.

The lithovore is a rock-eater about the size of a snowmobile but weighing three times as much. Its grasping tongs have the power of a pickaxe or a jackhammer, allowing it to break apart and pulverize chunks of crystalline rock for easier ingestion. Its highly specialized, multipurpose extendible proboscis, or nose, can make especially tough minerals more palatable by spraying the desired food source with a strong acid, which melts or dissolves it. The resulting fluid is then sucked into the body. Once inside, the material is applied to the lithovore's body for growth and repair.

The end of the proboscis also carries the equivalent of a Geiger counter, with which the creature locates radioactive minerals. These are an important energy source, since the lithovore "runs" on electricity generated by an internal reactor. Its limbs function like the high-powered hydraulic lifters seen on graders and cranes at a construction site. Lithovores are not born but are manufactured by a "queen," which looks more like a factory than a living creature. The queen builds small lithovores and releases them on this crystalline world, where they go happily about munching the rocks.

The dark, jagged object on the facing page is a giant comet—a huge flying mountain of ice that formed in the dark distant spaces surrounding a newborn star far smaller than our sun. The supercold ices of this comet, which have never risen more than a few degrees above absolute zero, contain abundant amounts of carbon. The carbon here, however, is an ultracold superconductor, so instead of producing organic carbon-based life, it acts as a nervous system. The entire comet could be a holistic consciousness. It may think. Yet even if humans were able to land on this living comet and explore it, we would probably not recognize it as a life form.

◆ The "lithovore," a hypothetical creature from a silicon-based biology, clambers across the crystalline "vegetation" on its home planet. Clanking along as it walks, the lithovore appears to us more like a mechanical creature than the animal life forms we are familiar with on Earth.

◆ A comet nucleus that has never sustained a temperature more than a few degrees above absolute zero is seen here silhouetted against a patch of celestial gas and dust. As explained in the text, an object like this could be capable of thought on some level, though humans would have great difficulty recognizing it as a form of life.

◆ The glistening surface of a neutron star is deceptively enticing. But this celestial body has three times the mass of the sun packed into an object as wide as a small city. Its surface gravity is 20,000 times stronger than that of Earth. Yet there could be life here—life as we don't know it.

Imagine a colossal sphere of matter so dense that a thimbleful of its material weighs a billion tonnes. Such objects exist. They are called neutron stars. They are born when a massive star dies, blasting itself to bits in a supernova explosion. The core of the star survives, imploded and crushed down to an object as wide as the city limits of a medium-sized town yet containing several times the mass of our sun.

A neutron star would look like a gigantic ball bearing, as seen in the illustration on the facing page. Its surface reflects the light of the stars of the Milky Way Galaxy, which look like a glittering necklace strung around the sphere.

The neutron star's surface seems almost inviting, like something that could easily be landed upon. But it is utterly alien and impossible to explore by humans. Compressing several times the sun's mass into such a small object produces immense surface gravity. A spaceship from Earth attempting to land on a neutron star would be crushed into a puddle of atomic particles.

On the surface of a neutron star, there are no atoms. They are ripped apart by high temperatures and the tremendous gravity. But there are many different kinds of atomic nuclei. The surface temperature of millions of degrees would allow those nuclei to combine to form macronuclei, perhaps like the large molecules that make up Earth life. Thus there could be creatures on neutron stars composed of macronuclei instead of molecules.

Macronuclei creatures would be so small that you would need a microscope to see them. But they could have all the complexity that we do. Their life functions would proceed at the speed of light, so a lifetime would be merely a fraction of a second. Every second of our time would span several generations on a neutron star. If neutron-star creatures visited Earth, they might not recognize us as a life form because we would appear frozen in time.

Perhaps the ultimate extraterrestrial life form would not be based on matter at all. Instead, it could be a pure-energy being, something which lives by absorbing energy and then using that energy for its thought processes and for communicating with the universe or with other pure-energy beings. But could we ever recognize such an entity? It might not appear much different from a star or the active core of a galaxy.

Would it recognize *us* as a form of life? This was the theme of the classic science fiction story *The Black Cloud* by astrophysicist Fred Hoyle. A mysterious cloud enters our solar system and begins to absorb energy from the sun. Meanwhile, Earth is being deprived of its normal flow of life-giving radiation. In the story, Earth scientists recognize that the cloud is, in fact, a thinking energy being which is completely unaware of our existence. The problem becomes how to communicate with something that is so alien and, without knowing it, so dangerous.

Life as we don't know it may exist, but we may never recognize it. Organic carbon-based life is the only life form that we understand. Just as we have difficulty guessing what form it might take, life as we don't know it may also have a difficult time recognizing us as living, conscious entities.

◆ A pure-energy being might be the ultimate alien form of life as we don't know it. But would we recognize it as life? Would it recognize us?

In 1967, astronomers at Cambridge University in England were trying out a new radio telescope that had been designed to scan wide areas of the sky and map low-temperature radiation from stars and gas in the Milky Way Galaxy. Jocelyn Bell, a student working at the telescope, reported an odd pulsing noise on the receiver. It was like the tick of a clock, very regular—bip, bip, bip—every 1.3 seconds. At first, the Cambridge astronomers thought it was interference from a nearby electric transformer or something else here on Earth. But they soon found that it was coming from a precise spot in the sky.

The astronomers timed the pulses for weeks and determined that they arrived *exactly* once every 1.33730109 seconds. That's more accurate than a clock which gains or loses just one second per year. Could this be a signal from another civilization?

While the scientists struggled to understand what they were observing, they playfully dubbed the signals LGM, for "little green men." Eventually, they realized that they were not receiving a greeting from little green men—or from anybody else. They had discovered the first pulsar, the dense core of a collapsed star that was spinning on its axis once every 1.33730109 seconds. Each time the pulsar spins, we receive a burst of radiation when it faces in our direction, like the rotating lamp of a lighthouse.

That is as close as we've come to detecting communications from extraterrestrials. There have been other false alarms, but none quite as dramatic. Now, several radio telescopes scan the sky 24 hours a day to search for alien signals. Hundreds of other radio telescopes study the natural radio radiations from stars and nebulas. None of them has picked up anything that could be considered a signal from another intelligence.

Meanwhile, we have been doing some signalling of our own. For nearly half a century, we have been unintentionally broadcasting our existence to the universe. We are a noisy species. Television and FM radio as well as military radar and other forms of communication signals easily leak into space. They have been hurtling outward at the speed of light for decades and now form an expanding bubble of babble that is about 50 light-years in radius. Within that distance of Earth, there are more than 1,000 stars.

It is unlikely, however, that extraterrestrials receiving these signals would be able to watch reruns of *Star Trek*. So many things are being broadcast simultaneously all over Earth that the signals escaping into space are probably jumbled to unintelligible static. But any extraterrestrial intelligence would know that the source of this type of racket could only be a planet with advanced technological life forms.

Other civilizations may be inadvertently announcing themselves to the universe just as we are. So far, we haven't detected any. Of course, extraterrestrials may have advanced to more efficient methods of

◆ Left: A robotic space probe reaches Alpha Centauri 86 years after it was launched from Earth in the year 2064. It becomes the first human-made device to reach another star system. After going into orbit around Alpha Centauri (the bright object at top), the probe will spend decades surveying every planet and moon in the system and transmitting a huge atlas of images back to Earth.

◆ Right: Several giant radio telescopes have been used to listen for signals from extraterrestrials that might be transmitting in the Earth's direction, either deliberately or by accident. So far, nothing has been detected.

communication that we would not recognize.

Because the distances between the stars are so immense, even radio signals, which travel at the speed of light, seem slow. Alpha Centauri, the nearest star, is 4.3 light-years from Earth. A message sent to someone in the vicinity of Alpha Centauri would take 4.3 years to get there. But that is speedy compared with travel time by spacecraft.

Today's fastest robotic space probes travel in the range of 100,000 kilometres per hour. At that speed, it would take about 40,000 years to reach Alpha Centauri. Nuclear-rocket propulsion could increase the speed by 10, or perhaps 100, times. And exotic propulsion systems of the distant future might whittle the travel time down further.

But there is a limit. No creature or object made of flesh or plastic or metal—anything that makes up humans and the tools we need—can exceed the velocity of light. Space warps, in the form of gravitational wormholes, may provide gateways to surpass this barrier, but ordinary matter would be ripped apart, atom by atom, on the way through.

Assuming that no profound breakthroughs are on the horizon, sending robotic probes to the stars at 1 or 2 percent of the velocity of light seems to be the next logical step. Small ships designed to repair themselves and to operate in space for hundreds of years could explore another star system for decades, sending back close-up pictures of every planet and moon. Robotic probes of this type might be sent to thousands of stars. Future generations would reap the bonanza of images and data flowing back in the centuries ahead.

This type of exploration seems unlikely to occur in today's political climate, where governments seldom think beyond the next election. But someday, it might happen. Only then will we know for sure what other planetary systems are like, because no matter how powerful telescopes become, there is no substitute for actually going there.

◆ The bright star near the centre of this photograph is Alpha Centauri, the nearest star to our solar system. It is 4.3 light-years, or 40 trillion kilometres, from Earth. Every other star in the picture is farther away. Some are more than 4,000 light-years distant (one light-year, the distance light travels in a year, is 10 trillion kilometres).

"Where are they?" It is a simple question, but it sums up the mystery of extraterrestrials. If they are out there, why haven't we been contacted? Why is there no evidence of their existence?

There are many possible reasons. Here are a few:

• We are the most advanced civilization in our sector of the universe; therefore, *we* have to find *them*.

• They are trying to contact us, but we have yet to notice.

• We are the least advanced civilization in our sector of the universe; therefore, they take no interest in us.

• They know about us but are leaving us alone as a galactic nature preserve.

• They have been studying us but do it in ways we cannot observe.

• Life was seeded on Earth by extraterres-trials in the far distant past, and we are simply an experiment in a galactic laboratory.

• We are alone.

If contact does come, there are four ways in which it might be likely to occur:

1. Extraterrestrials land here and show themselves or their equipment.
2. We receive some kind of signal from afar, perhaps by radio telescope.
3. We discover something they accidentally or intentionally left behind (on Earth or somewhere else).
4. We begin exploring the galaxy and dis-cover other life somewhere out there.

Contact scenario #1 is dramatically illus-trated at right, when a colossal alien ship hovers over a major city long enough for everybody to get a good look. Of course, *real* extraterrestrials will probably think of a far more ingenious method of introducing them-

selves. (For an exciting story about exactly that, try the classic science fiction novel *Childhood's End* by Arthur C. Clarke.)

In contact scenario #3, illustrated on this page, 21st-century explorers have just arrived on Jupiter's moon Callisto to inspect a clearly alien artifact that was first detected by the high-resolution cameras of a robot spacecraft in orbit around Jupiter. It is an eight-sided diamond shape, with one side resting on the surface.

The extraterrestrial intelligence that left this artifact on Callisto probably selected Jupiter's outer large moon because of its sta-ble orbit around the dominant planet in our solar system. Also, Callisto's heavily cratered surface is ancient and has apparently not been torn up by volcanoes or periodically melted, as are the moons closer to Jupiter. These factors ensure that the object would likely remain undisturbed for millions, perhaps billions, of years.

The artifact we see here is made of carbon pressurized to form diamond crystal, which is exceptionally durable over time and resis-tant to the ceaseless, erosive rain of micro-meteoroids. It is pockmarked by thousands of meteorite impacts, which give it an esti-mated age of one billion years. The most interesting question, of course, is: What did the advanced civilization hide within the gigantic diamond?

The discovery of such a clearly constructed artifact would tell us that intelligences superior to humans visited or inhabited our solar system long ago. How would people react to this news? At one time, such a reve-lation might have been too much for human-ity to handle. That is certainly not the case today. Most people already assume that extraterrestrials exist and would readily ac-cept such news. After a few days, the excite-ment would die down, and folks would go about their business as before—except for one thing: If extraterrestrials visited our solar system once, maybe they'll be back.

◆ Members of a future expedition from Earth approach a huge alien artifact—a giant diamond—on the surface of Jupiter's moon Callisto. This hypothetical scene depicts one of four ways that we might come into contact with extraterrestrials. A sec-ond one is shown on the facing page.

◆ Will extraterrestrials ever make deliberate contact with Earthlings? There are many possible explanations for why we have not been contacted yet. We may be alone in our part of the universe. We may be too primitive to be of interest to extraterrestrials. And consider what purpose such an event would serve. If extraterrestrials know about us, why would they disturb us? We might be far more interesting to them if we remain as we are.

The first person to write about extraterrestrials was Metrodorus, a Greek teacher and philosopher. Sometime around the year 350 B.C., he wrote: "To consider Earth the only populated world in infinite space is as absurd as to assert that in an entire field sown with seed, only one will grow." Much has been written since. Here's some of the best of it.

Science Fact

The Search for Extraterrestrial Intelligence by Thomas R. McDonough (Wiley, 1987). The best nontechnical book about the quest for extraterrestrials. Factual and entertaining.

Life Search by the editors of Time-Life Books (1988). This superbly illustrated book is part of Time-Life's excellent "Voyage Through the Universe" series. Highly recommended.

Life in Darwin's Universe by Gene Bylinsky (Doubleday, 1981). Examines the possibilities of life in the universe and suggests various forms extraterrestrials might take.

Life Beyond Earth by Gerald Feinberg and Robert Shapiro (Morrow, 1980). A large book (464 pages) that goes into detail. Worth the effort to find and read.

First Contact edited by Ben Bova and Byron Preiss (Penguin/Plume, 1990). A collection of 21 essays about extraterrestrial life by Isaac Asimov, Arthur C. Clarke, Gregory Benford and other top writers.

The Universe and Beyond by Terence Dickinson (Camden House, revised edition, 1992). A heavily illustrated tour of the universe. Contains a detailed discussion of the search for extraterrestrial life.

"Science Goes to the Movies" by Jack Weyland (article in *Science Year 1994*, supplement to the World Book encyclopedia). Explores the scientific blunders in film and television science fiction.

Science Fiction

There are hundreds of science fiction novels about contact with extraterrestrials. Some of the best are *The Star Beast* by Robert Heinlein, *Catseye* by Andre Norton, *The Left Hand of Darkness* by Ursula LeGuin, *Deathworld* by Harry Harrison, *Chocky* by John Wyndham, *Mission of Gravity* by Hal Clement, *Empire Star* by Samuel R. Delany, *Hospital Station* by James White, *The Black Cloud* by Fred Hoyle and *The Gods Themselves* by Isaac Asimov. Two classics of extraterrestrial contact are *The War of the Worlds* by H.G. Wells and *Childhood's End* by Arthur C. Clarke.

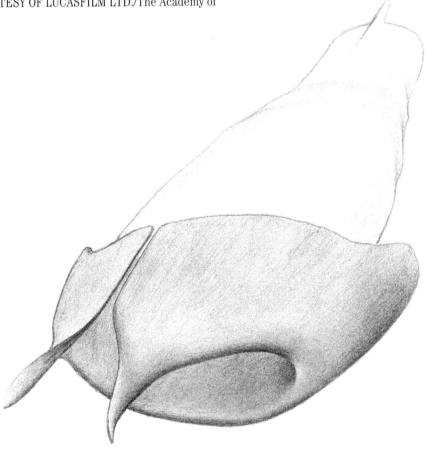

Terence Dickinson

Terence Dickinson has been writing about astronomy and space exploration since the 1960s. He is the author of 11 books and more than 1,000 magazine and newspaper articles. In 1974, when he was editor of *Astronomy* magazine, he was visited by a young artist, Adolf Schaller, who offered a painting of Saturn for use in the magazine. Dickinson used the painting in the next issue and has been a fan of Schaller's work ever since. They have collaborated on several books, but this is their most extensive cooperative venture. Dickinson has received numerous national and international awards for his work, among them the New York Academy of Sciences Book of the Year award and the Royal Canadian Institute's Sandford Fleming Medal for achievements in communicating science to the public. In 1994, the International Astronomical Union officially named asteroid 5272 "Dickinson."

Adolf Schaller

Adolf Schaller is an artist, writer, poet and composer whose work reflects his passion for science and nature. He is best known for his astronomical and science-related illustrations, which are noted for vivid realism and scientific accuracy. His work has appeared in many published forms, from educational slide sets, filmstrips and posters to numerous magazines and books, including *The Universe and Beyond* by Terence Dickinson and *Cosmos* by Carl Sagan. He has also worked in the television and motion-picture industries as a visual-effects artist, supervisor, technical consultant and science advisor. In 1981, he received the prime-time Emmy Award for outstanding individual achievement in creative technical crafts for his work on the PBS science series *Cosmos*, hosted by Carl Sagan.